Cerebrum 2008

Cerebrum 2008

EMERGING IDEAS IN BRAIN SCIENCE

Cynthia A. Read, Editor

DANA
PRESS

New York • Washington, DC

The Dana Foundation
745 Fifth Avenue, Suite 900
New York, NY 10151

900 15th Street NW
Washington, D.C. 20005

DANA is a federally registered trademark.

ISBN: 978-1-932594-33-1

ISSN: 1524-6205

Cover design by Tobias' Outerwear for Books
Cover illustration by Brand X Pictures

Contents

Foreword

by Carl Zimmer

Carl Zimmer is a science writer and author of *Soul Made Flesh* (Free Press, 2004), a history of the brain, which was named one of the top 100 books of 2004 by the *New York Times Book Review*. He won a 2007 Communication Award from the National Academies. His other books include *Evolution: The Triumph of an Idea, At the Water's Edge,* and *Smithsonian Intimate Guide to Human Origins.* Zimmer is a frequent contributor to the *New York Times, National Geographic, Discover,* and *Scientific American.*

NEUROSCIENCE IS PERHAPS the most obsessed-over branch of all the sciences. As a rule, the public and the pundits don't care much about science. You do not hear drive-time talk show hosts taking calls about the latest studies on soil mechanics or quantum physics. But when the study in question deals with the human brain, the normal rules no longer apply. Much of today's science news—not to mention talk shows and conversations around the water cooler or the blogosphere— deals with the brain. The news may be about a new drug for strokes, for example, or a new study on autism, or an electrode that restores consciousness. We want it all.

The trouble is, we don't know what to do with it. Part of our trouble is the sheer mass of data that's bowling us over. As a science writer, I have gone to many scientific conferences, but I shudder at the thought of ever attending the Society for Neuroscience's annual meeting. In recent years, each conference has offered well over 10,000 abstracts.

How could I possibly get any sort of understanding of where the field is as a whole?

We also don't know what to do with all the news about the brain because each scientific report is just one minuscule jigsaw puzzle piece. Scientists can spend an entire career on a single nook of the brain, a single type of receptor. Most of the jigsaw is still missing, and when neuroscientists themselves try to come up with overarching hypotheses to make sense of their results, it turns out that the pieces can fit together in many different ways.

The third part of our trouble, I believe, has something to do with the way our minds work. A great deal of research indicates that we are constantly filling in gaps, constructing coherent pictures of our world. Those pictures are not just of landscapes or faces, but also of human nature itself, of the moral universe. We use new information to fill in those pictures, but often we distort that information to fit. If it does not fit, we reject it altogether. Conservatives are not the only ones to do so, I should point out; neuroscientists who study the biological foundations of intelligence sometimes find themselves under attack from the left, as if they were Nazis for simply wanting to understand cognition.

This is not a new phenomenon. It existed more than three centuries ago, when neurology was first coming into existence. You can see it on display in the writings of Thomas Willis, a seventeenth-century physician who is considered the father of neurology. He carried out extraordinary studies of the brain, recognizing it as the seat of thought, emotions, memories, and the rest of what we think of as ourselves, producing all these faculties through the same kind of reactions as those that make bread rise or gunpowder explode. And yet Willis was not immune to the urge to fill in the gaps. As a loyal soldier of King Charles I, he saw his mission to be healing the madness of the English Civil War and the excesses of Puritans. As a wealthy seventeenth-century Englishman, he believed that all peasants were, to use modern language, mentally disabled, and that he had discovered why: the pathways in peasant brains were obstructed, so their animal spirits traveled too slowly.

While these habits may have a deep history, we need to overcome them. Our collective decisions will depend more and more in the future on the insights of neuroscience. We need to find a way to appreciate these insights, and learn how to apply them to our own lives. This book shows how we can do that. It gathers a series of provocative articles and reviews published over the past year in the online journal *Cerebrum*. Together, they offer a guide to ordering one's understanding of the brain.

Much of the work in neuroscience these days is dedicated to understanding how the brain goes awry, through either infection or dysfunction. Scott P. Edwards surveys recent research on some of the most enigmatic molecules, prions. These misfolded proteins are the cause of mad cow disease and several other emerging disorders. Edwards examines the struggle to understand how prions malfunction and how to translate that knowledge into new treatments. While prions are a recently recognized cause of brain disease, cerebral malaria is an ancient scourge. Nevertheless, it still manages to kill hundreds of thousands of people a year, many of them children. Kayt Sukel explains new insights that are emerging about how the *Plasmodium falciparum* parasite wreaks its havoc on the brain. And Vivian Teichberg and Luba Vikhanski look at how the brain's own supply of chemicals can devastate the organ when a wave of the neurotransmitter glutamate is unleashed during a stroke. Many labs are searching for medical treatments for these and the many other brain disorders. E. Ray Dorsey, Philip Vitticore, and Hamilton Moses describe a promising new class of drugs, known as "biologics," that are based on living organisms and may be able to treat brain disorders that have previously been untreatable.

What makes neuroscience a particularly rich science is that it examines an organ that must be understood on many levels at once. At the base of this hierarchy are neurotransmitters, receptors, and other features of individual neurons. But these components of the brain allow higher levels of the brain's hierarchy to emerge. Michael J. Frank explains how the neurotransmitter dopamine allows a circuit of neurons in the basal ganglia to control our subconscious habits of thought and action. Karl K. Szpunar and Kathleen B. McDermott investigate the circuits that enable us to remember the past, thus allowing us to plan for the future. Because these circuits are complex, they can be very difficult to study. Some of the best clues to how they work come when their parts malfunction. In "When Music Stops Making Sense," Petr Janata looks at how strokes and other brain injuries help to reveal the nature of the brain networks that allow us to appreciate music. When these sorts of injuries occur, the brain sometimes displays a wonderful capacity for recovery, a capacity that Michael E. Selzer describes in his article on rehabilitation after brain injury.

The hierarchy of neuroscience actually transcends the brain itself. That's because a human brain is embedded in an environment that influences it, and that it influences in turn. The stress that we experience within our brains, explains Fabienne Mackay, can affect how well our immune systems fend off infections. Treating Alzheimer's disease is not just a matter of developing better drugs, as Kayt Sukel and Russell Epstein explain. By appreciating how we perceive and relate to the spaces around us, architects can help make the lives of people with Alzheimer's easier.

Neuroscience is also inseparable from ethics. It raises many difficult questions, two of which are specifically addressed in this volume. Anjan Chatterjee explores the issue of when it is right to let people use a drug to erase their painful memories, and Mark Hallett and Paul R. McHugh debate whether free will even exists. Even politics is not separate from neuroscience. After all, our political views and decisions are shaped within our brains. Instead of probing the differences between liberal and conservative minds, David A. Drachman ponders the very practical question of how to decide who is mentally fit to vote.

You will find much to contemplate in this book. But that does not mean you are about to be swept up into a cyclone of chatter about the brain. Here, instead, the complexity of the brain gives rise to a stirring symphony dedicated to this most wonderful organ.

CHAPTER 1

Building for the Shattered Mind

Partnering Brain Science and Architecture

by Kayt Sukel and Russell Epstein, Ph.D.

Kayt Sukel is a writer whose essays and articles have appeared in *Science, Memory and Cognition,* and *NeuroImage,* as well as the *Washington Post,* the *Christian Science Monitor,* and *National Geographic Traveler.* She can be reached at ksukel@hotmail.com.

Russell Epstein, Ph.D., is an assistant professor at the Center for Cognitive Neuroscience and the Department of Psychology at the University of Pennsylvania. His research focuses on how the brain represents scenes, places, objects, and events, and how it uses these representations to support spatial navigation and action. He can be reached at epstein@psych.upenn.edu.

What might architects learn from neuroscience that would help in designing better nursing homes and other facilities for the aging, particularly people with Alzheimer's disease? When the brain is impaired, the environment can make it easier or harder to find one's way, remember habits such as how to get dressed, interact socially, eat enough to be nourished, and even walk safely outside. Coupled with existing knowledge about the brain, answers to research questions identified by participants in a recent interdisciplinary workshop should one day guide the design of facilities that will improve life for people with Alzheimer's and perhaps for all of us as we age.

IN *ELEGY FOR IRIS*, John Bayley's poignant memoir chronicling life with his wife, Iris Murdoch, as she struggled with Alzheimer's disease, the author writes, "Alzheimer's is, in fact, like an insidious fog, barely noticeable until everything around has disappeared. After that, it is no longer possible to believe that a world outside the fog exists."

Alzheimer's disease, a progressive and irreversible neurodegenerative brain disorder, currently affects more than 5 million Americans. The disease causes formation of plaques in the brain's cortex and leads to degeneration of neurons, a reduction in key neurotransmitters such as acetylcholine, serotonin, and norepinephrine, and a loss of synaptic activity—the means by which neurons communicate. This brain atrophy causes symptoms that begin with simple memory loss and gradually advance to widespread, persistent cognitive impairment that may include problems with critical reasoning and sensory perception, general confusion, and social withdrawal. Because of this overwhelming loss of function, people with the disease will eventually need round-the-clock care, often provided by nursing homes or assisted-living facilities. Even the best of these facilities, however, can be a tremendous adjustment for people struggling with the disease, as well as for their families.

But what if we could create assisted-living spaces for people with Alzheimer's that could make life easier despite the "insidious fog"? What if, by bringing together knowledge of architectural design and knowledge of what goes on inside the brain of the person with Alzheimer's, we could design buildings and interiors that would help people stay more capable over longer stretches of time, remember the outside world, and successfully interact with it? Furthermore, what kind of brain research is needed as a basis for creating facilities that would make the adjustment from home to a care facility less stressful for the patient? To begin asking the questions that could jump-start this process, the Academy of Neuroscience for Architecture (ANFA) held an interdisciplinary workshop, "Neuroscience of Facilities for the Aging and People with Alzheimer's," in late November 2006.

"Right now, not very much knowledge is available in neuroscience that is applicable to the design of facilities for people with Alzheimer's," says John P. Eberhard, FAIA, founding president of ANFA. "Ninety-nine percent of research in neuroscience is oriented towards disease and the ramifications of that disease. Practically no one looks at research that could be used to improve the facilities such people live in. We're trying to encourage that to happen."

Over three days in Washington, DC, two dozen prominent neuroscientists, architects, and experts on Alzheimer's disease and aging came together to discuss how to promote neuroscience research that will eventually have tangible application to designing nursing and assisted-living facilities. In this article, we focus mainly on Alzheimer's disease, but the lessons learned with this disease can be applied to the aging population as a whole.

"The purpose of the workshop was to create hypotheses that potentially can be tested in neuroscience laboratory conditions and eventually yield results that architects can use in design," says Eberhard. To that end, participants formed small working groups focused on specific issues, among them physiology and physical ability, memory, sensory perception, and cognitive mapping. Each working group developed a series of hypotheses that, if tested, might have great impact on how architects and designers determine design criteria for assisted-living spaces in the future.

Preserving Memory

Although Alzheimer's disease affects many facets of cognition, memory is the ability that is first noticeably diminished. People with Alzheimer's often have difficulty placing names and words and recalling events, as well as problems with planning and organization. The workshop's memory working group formulated several hypotheses that the members believe can help provide design guidelines for future Alzheimer's facilities. Most of these hypotheses had to do with understanding how the human brain, both the normal brain and the brain affected by disease, uses sensory information to cue memory retrieval.

Orienting Toward Activity

How does the brain recognize information out in the world and figure out what to do with it? Can the presence of immediately accessible visual information help people know where they are and orient them toward activities that are appropriate? Members of this working group would like to find out whether direct visual contact can bring about desired activities. For example, having a toilet in direct line of sight might cue people with Alzheimer's to use it and thereby avoid accidents. Similarly, having a window in the room might help orient people to the time of day and appropriate activities, such as getting dressed in the morning or going to the dining area for meals. An understanding of how the brain recognizes objects and applies that recognition to their use would be a great benefit in design.

Engaging Procedural Memory

The group hypothesized that providing culturally relevant activities and spaces can increase social engagement and improve quality of life. The members believe that lifelong activities such as family meals and social events may have become hardwired in the brain as habitual procedures. These procedural memories might be stored in brain regions such as the basal ganglia, which research has shown is a key structure for encoding habits. The basal ganglia are less affected by Alzheimer's disease than are regions such as the hippocampus, which supports memories for facts and events. Therefore, the right cues and activities might help people with Alzheimer's to more easily retrieve those habits and behavioral routines, even if retrieval of other kinds of memory is impaired. The ability to read words or simple phrases appears to be retained far into the course of dementia, which suggests that signage might provide a useful cue for the retrieval of procedural memories. If the hypothesis that procedural memories are preserved is borne out, architects can incorporate this knowledge into design of spaces that integrate the right kinds of visual and other sensory cues to help people with Alzheimer's retrieve these memories and engage more in daily activities.

Finding the Way

People with Alzheimer's disease often lose the ability to find their way. This critical skill requires a great deal of what is called executive function, such as planning and selective attention, as well as the ability to coordinate and direct one's movements. As the prefrontal cortex deteriorates, however, executive function becomes diminished and such tasks cannot be performed as usual.

The working group hypothesized that specific color and space cues can help people differentiate between personal and public space, as well as help them successfully navigate within a facility. The group would like to see neuroscientists test what kinds of visual cues, as well as cues for other sensory modalities, such as sound and touch, are most salient to people with Alzheimer's disease, as well as how the cues can best be utilized to help with spatial navigation.

A Little Daylight

The memory group was also interested in the source and amount of light in facilities. A 1999 study by an architectural consulting firm, the Heschong Mahone Group, showed that students who took classes in rooms with more natural light performed better on standardized tests than those whose classrooms were illuminated by artificial light. It is hypothesized that one of the ways daylight helps learning is by improving memory. Why is this? The members of the working group would like to understand the mechanisms underlying this memory enhancement by daylight and how it might be applied in designing facilities for people with Alzheimer's.

With better understanding of how memory is cued by the environment in all of these areas, architects may one day be able to use the information to design better environments that make memory cues more noticeable and relevant to people in the various stages of Alzheimer's.

The Five Senses

As Alzheimer's progresses into its later stages, people with the disease often develop agnosia, or the loss of ability to identify common objects by sight. Those with this syndrome are not blind—often they can copy drawings of the very objects they cannot identify and can report purely visual information about these objects (such as their perceived color). Nor have they lost their semantic knowledge about objects; when they hear the name of an object, they can give an appropriate definition. Rather, agnosics suffer damage to the high-level visual representations in the occipital and temporal cortices that normally mediate our recognition of objects by sight without effort. Interestingly, however, the other senses can sometimes pitch in to help agnosics recognize objects. By touching the object or hearing it make a sound, people with agnosia can often recognize the same item they were unable to identify by relying on visual input alone.

If environments can be designed to provide additional cues for object recognition, people with Alzheimer's may be able to act independently for longer periods of time. For example, if a person is unable to recognize a shoe by sight, designing closet spaces that place footwear off the floor and within easy reach may encourage touching the shoes and thereby may facilitate identification. Paradoxically, however, too much incoming information has the potential to overwhelm these patients. The workshop participants who focused on sensory perception came up with two sets of questions for neuroscientists that explore this tension.

Is More Really More?

In an attempt to help activate sensory areas of the brain, designers of many facilities for Alzheimer's patients have assumed that more sensory information is better. Items such as patterned carpets and brightly colored decorations have been recommended for use without empirical justification. But is more information really helpful to persons with Alzheimer's?

Although neuropsychological research has shown that multisensory cues aid performance of perceptual tasks in healthy individuals, the situation with people with Alzheimer's may be different. The disease produces a progressive degeneration of the senses, so people with the disease may be able to process sensory information more easily if they encounter less complexity in their environment. Members of the sensory perception working group would like researchers to examine a series of specific questions related to sensory cues. Do patterned carpets produce more confusion than single-colored ones? Are multiple signs or cues that help direct people to a destination more helpful than single cues? Which sensory cues—visual, auditory, olfactory, touch—are most effective for people with Alzheimer's disease and how might they be best combined to promote recognition?

In addition, Alzheimer's specialists have had success in using behavioral conditioning methods to help people with Alzheimer's relearn how to perform certain activities. For example, some facilities help residents remember to go to the dining area by playing the same song over the public announcement system at lunchtime each day or training residents to use index cards as cues. In effect, these conditioning methods help people with Alzheimer's remember information by activating pathways in the brain that have not yet been ravaged by the disease. But facility designers and managers would like to better understand the mechanisms underlying this kind of learning so they can use it with residents more effectively. Are multiple simultaneous cues more effective for patients? Or are they in fact more confusing? Are spatially localized or ambient cues more likely to facilitate this relearning process? And what is the relationship between the two? By tackling these questions, neuroscientists can help architects balance the benefits of minimizing architectural complexity (thus making the environment simpler for the residents to deal with) with the benefits of providing sensory cues for the residents (thus resulting in a richer environment with more ways for residents to gain understanding).

Providing Kitchen-ness

What makes a room a specific type of room? For example, what makes a kitchen a kitchen? Is it the sight of a table and chairs? The hum of the appliances? The smell of fresh-baked bread? The taste of orange juice? What sensory cues can we provide people with Alzheimer's disease so that they can recognize the category of "kitchen" and retrieve the correct social schemas to know what to do there? Researchers call this kind of knowledge "semantic memory." Though people with Alzheimer's disease have some semantic memory impairments, semantic memory tends to be less impaired than spatial or episodic memory. If the right number and type of recognition cues are incorporated into the design of an environment, persons with Alzheimer's may be able to more easily retrieve memory of semantic categories, like kitchen-ness, and thus be encouraged to engage in the behaviors appropriate to that room.

All of these questions have a profound impact on design. For example, many assisted-living facilities market small apartments to help residents live more independently. However, it is possible that studio apartments that attempt to include facets of several different rooms in one space (for example, a half kitchen on the wall of a living room), which are often very different from a resident's previous home, may cause more confusion for residents in the long run. Greater education on how Alzheimer's affects semantic memory has the potential to provide designers with the ability to create spaces that will be readily recognized and useful to these patients.

Room to Move

In people with Alzheimer's, decline of physical abilities usually begins with minor problems in coordinating movement as a result of deterioration in the frontal and motor cortex. As the disease progresses, however, physical abilities degenerate to the point where people require assistance with simple, everyday tasks such as getting dressed or feeding themselves. This physical deterioration is closely linked with other physiological symptoms, including weight loss, sleep disruption, and loss of muscle control. Members of the physiological and physical ability working group came up with four hypotheses that might be tested.

Let There Be Light

Like the memory group, the physiological and physical ability working group was interested in light. As Alzheimer's advances, weight loss becomes a serious issue. Brain deterioration in areas such as the orbitofrontal and occipital cortices makes it more difficult for patients to interpret social, physiological, and other cues reminding them to eat. How might this problem be postponed or averted? One possibility is increasing the level of light in dining areas. The thinking is that brighter lights may allow people with Alzheimer's, who often suffer visual impairment, to see better and therefore to obtain visual cues that will stimulate appetite. Members of the group hypothesized that increased visual contrast achieved by brighter lighting would allow patients to see not only the food itself but also other people around them who are eating, thus evoking additional social cues that stimulate hunger.

The Great Outdoors

Increased physical activity has long been known to enhance overall health and increase life expectancy. However, along with the declining mental and physical abilities of people with Alzheimer's comes less spontaneous physical activity. This problem is exacerbated because of safety concerns such as the danger of falls or that an unaccompanied patient might wander off the premises, and consequently many assisted-living facilities do not provide outdoor spaces for residents.

The Woodside Place Alzheimer's Residence in Oakmont, Pennsylvania, however, is one facility mentioned at the workshop that has had great success with allowing the residents to have access to the outdoors. Outside, a system of looping paths was designed so that residents can safely wander on the grounds. And residents are taking advantage of this, frequently spontaneously deciding to take a walk outside.

But this success leads to several questions that need to be answered scientifically. Do people have an innate need to be outside or to explore? How can we help motivate people with Alzheimer's to increase their physical activity? The physiological and physical ability working group would like to see such questions tested in people with Alzheimer's to find out if these needs exist, and if so, whether they increase or decrease as the disease progresses. Greater understanding may help architects to design outdoor areas that not only engage the interest of people with Alzheimer's but also are safe enough for continued use as the disease progresses.

Pets and Children

Pets have long been thought to benefit the health of elderly people in retirement communities or facilities. Several studies have shown that the presence of a pet can reduce stress, lower blood pressure and cholesterol, and increase physical activity. Those benefits also apply to persons with Alzheimer's disease. In a landmark study in 1989, Jacqueline Stolley of the Trinity College of Nursing and Health Sciences and colleagues found that the presence of a dog significantly increased the number of social behaviors exhibited by a group of twelve institutionalized people with Alzheimer's. This finding has been replicated several times in the past two decades, and as a result, many long-term-care facilities have resident or visiting pets available as "pet therapy."

Similarly, anecdotes from caregivers, as well as empirical research, have shown that having children around can benefit people with Alzheimer's. Several studies have demonstrated that intergenerational activities that include the elderly and children increase social behaviors and improve affect in people with dementia. But why is this so? Why are animals and children able to reach people with dementia when adults are not? Might it have something to do with the neural deterioration that occurs in people with Alzheimer's, often reducing them to a childlike state? Or is it that dogs and children provide such clear social cues that they are obvious even to people in the advanced stages of the disease? To date, the benefits of these kinds of therapies have not been studied from a neuroscientific perspective.

Although facilitating activities involving children or pet therapy in a nursing home may seem to be an operational issue, it is not without architectural impact. Greater understanding of the brain basis for the benefits of these activities could be leveraged to design buildings that facilitate introducing children and pets, both for their effects on physical health and as cues to evoke social behavior. Architects and designers might consider, for example, providing easier access to the outdoors, concealing potentially intimidating medical equipment, and creating pet- and child-friendly spaces within assisted-living facilities that are safe and comfortable both for residents and for children and animals.

The Magic Number

Residents in long-term-care facilities are often put into groups to stimulate physical activity and social interaction. But how many people should be in these groups to give the most benefit? Members of the physiology and physical abilities working group hypothesized that smaller groups (for example, nine people) would elicit greater benefits than larger groups (of fifteen or more). They suggested that the greater memory demands necessary for understanding the social relationships within the larger groups might be overwhelming to people with Alzheimer's. Smaller groups may be less confusing, allowing people with Alzheimer's to be able to interact with one another better.

When regulations or other imperatives necessitate large-group activities, it may be possible to address this issue by arranging multiple smaller groups to be conducted at the same time. Pilot research has even shown that people with early dementia are capable of successfully leading small-group activities for those with more-advanced dementia, making this "multiple small groups" approach more feasible. Workshop participants suggested that scientists examine group size to measure the effect on social behavior and other abilities. The knowledge could be applied to planning social therapy and also to the design of spaces for social activities.

Cognitive Mapping and Design

The ability to get from point A to point B is a critical cognitive skill. Wayfinding was important enough to garner its own working group but was also of great interest to those in the memory and sensory perception working groups. The seemingly simple act of finding one's way comprises many smaller tasks, such as knowing where we want to go, understanding how to get there, having the ability to coordinate our movements in order to physically make our way to the desired destination, and recognizing the destination once we reach it. All of these aspects of wayfinding can be severely diminished in people with Alzheimer's disease. Workshop members concerned with cognitive mapping—the ability to internally represent the spatial relationships between different locations in the environment—discussed how to make environments easier to interpret.

The Impact of Legible Environments

Once again, discussion came back around to salient cues. Working group participants hypothesized that environments that are easier to interpret ("legible") would be easier to navigate, thus increasing social engagement and affect and decreasing disruptive behavior in people with Alzheimer's disease. But before this specific hypothesis can be tested, neuroscientists must be able to define just what sensory information is legible to those with the disease. A greater understanding of the types of cues that facilitate learning and relearning, memory, and wayfinding is necessary before architects and designers can create environments that are easier for people with Alzheimer's to interpret.

Balancing Challenges and Abilities

The late M. Powell Lawton, Ph.D., a pioneering researcher in the psychological and social aspects of aging, identified the construct of "environmental press," defining it as the challenges existing in an environment that can make activities difficult. These challenges must always be balanced with the coping abilities of the people for whom one is designing. For example, something as simple as a short staircase may offer too great an environmental press for residents of an assisted-living facility, given their declining movement coordination and physical strength. The idea is that a well-designed facility reduces environmental press in such a way as to improve the daily quality of life.

But defining environmental press for Alzheimer's disease can be difficult, because the disease progresses in stages and along a very individualized timeline. Each person with the disease will have different levels of ability and different challenges to overcome at different times, making it difficult to apply generalized strategies for reducing environmental press.

Many aspects of facility design affect the environmental press, among them the layout and design of rooms, grounds, and common areas. The working group focused on social interaction as the primary driver for reducing challenges for people with Alzheimer's disease. In doing so, they came up with two hypotheses they would like to see tested.

Dining in Style

Nutrition and caloric intake, a serious issue in the care of people with Alzheimer's disease, was discussed not only by the physiology and physical ability group but by the environmental press working group as well. This group wanted to know if the way people dine affects how much they eat and therefore increases or decreases their overall level of health. Members of the group were interested in comparing family-style and restaurant-style dining, defining "restaurant style" as two to four people who are served by waitstaff and "family style" as a larger group of eight to ten people who pass around large platters of food and serve themselves. The working group hypothesized that family-style dining might promote better socialization, eating habits, and overall health.

Social Interactions for Health

How do the behaviors that make up social interaction affect health? Evidence has shown that people who frequently engage in social activity tend to be healthier than those who do not. Given this information, the working group also hypothesized that settings that foster social interactions among large groups of people are associated with better health outcomes. This was in direct contrast with the physiological and physical ability group, who felt that smaller social groups would be more beneficial. Clearly, optimal group size for people with Alzheimer's is an area that needs further research.

From a neuroscience perspective, social interaction is a compilation of many complex behaviors that are difficult to tease apart. But perhaps greater understanding of the neuroscience of social behavior and social interaction can lead to a model of how the neural systems that mediate these behaviors decline with the progress of Alzheimer's. Such a model would provide evidence-based design criteria for many different aspects of care for residents of assisted-living facilities.

Paving the Way

More than a hundred years ago, before doctors had an understanding of germs and how they were spread, epidemics of disease ran rampant through hospitals. Once germs were discovered, however, and medical and biological researchers worked together to determine how they were spread, changes were made. Nurses began washing their hands. Medical waste was properly disposed of. Antiseptics were introduced. And the rate of infection decreased in response to these measures. By working together in a cooperative, interdisciplinary fashion, medical doctors, biologists, and other critical groups of interest were able to change the way patients were treated and, in turn, improve their quality, and even their length, of life.

ANFA hopes that a similar shift will occur in the worlds of architecture and neuroscience. "The interface of architecture and neuroscience is fertile ground for innovation. I think there will be some interesting collaborative work that proceeds from this workshop," says workshop participant Cameron J. Camp, Ph.D., director and senior research scientist at the Myers Research Institute in Ohio.

As more professionals realize the importance of knowledge-driven and empirically based design, experts from the fields of neuroscience, architecture, and Alzheimer's research can join forces to formulate the right questions—and, in time, discover their answers—to help construct long-term-care facilities that will improve the quality of life for people living in the insidious fog of Alzheimer's disease, as well as for their aging peers.

And given the growth in the numbers of people living well into advanced age, coupled with the resulting exponential increase in Alzheimer's disease, this collaborative effort is something from which we may all one day benefit.

CHAPTER 2

Remembering the Past to Imagine the Future

by Karl K. Szpunar and
Kathleen B. McDermott, Ph.D.

Karl K. Szpunar, who joined the Memory and Cognition Laboratory at Washington University in St. Louis in 2003, is using behavioral and functional neuroimaging techniques to research human memory. Before coming to Washington University, he worked with Glenn Schellenberg, Ph.D., and Endel Tulving, Ph.D., at the University of Toronto. He can be reached at karl.szpunar@wustl.edu.

Kathleen B. McDermott, Ph.D., is an associate professor in the Departments of Psychology and Radiology at Washington University in St. Louis and principal investigator of the Memory and Cognition Laboratory. Her specialty is in the area of human memory, which she investigates using behavioral and functional neuroimaging techniques. She can be reached at kathleen.mcdermott@wustl.edu.

Remembering experiences in our past and imagining ourselves in some future event both involve a kind of mental "time travel." Now neuroimaging has shown how these processes are intrinsically linked in our brains. The authors believe that learning about this connection may help us better understand disease in which memory is affected, as well as basic human traits such as motivation and creativity.

REMEMBER FOR A MOMENT a recent gathering with friends. Who was there? When and where did the event take place? Now imagine a future gathering with friends. Who is there? When and where does the event take place? As with mentally reliving the past, mental representations of the future include images of personal episodes taking place in specific settings.

According to cognitive psychologist Endel Tulving, Ph.D., the ability to envision specific future scenarios (called episodic future thought) may be closely related to the ability to recollect specific episodes from our past (that is, episodic memory). Indeed, evidence from neuropsychology, clinical psychology, and developmental psychology indicates that people who cannot remember specific details from their past also appear to be impaired in their ability to mentally envision personal future experiences.

Efforts to understand the relationship between memory and future thought are in their infancy. The potential payoff, however, is considerable: it will likely inform our basic understanding of human memory, as well as of brain disorders in which that most essential human ability is impaired.

Evidence from Damaged Brains

Consider the patient known in the scientific literature as KC. KC, who has been studied extensively by Tulving and his colleagues at the

University of Toronto,[1] has global amnesia caused by diffuse brain damage that he sustained in a motorcycle accident. Many of KC's cognitive abilities are intact, but he can neither remember any single episode from his past nor project himself mentally into the future. When asked to do either, he states that his mind is "blank"; when asked to compare the kinds of blankness in the two situations, he says it is the "same kind of blankness."

DB, another brain-injured patient who has been studied by Stanley Klein, Ph.D., and his colleagues at the University of California at Santa Barbara, exhibits a similar profile. Following DB's cardiac arrest, his brain sustained damage due to lack of oxygen. DB can no longer recollect his past, nor can he project himself into the future.[2] Both KC and DB have self-concepts consistent with descriptions of their personalities given by others who know them well. Although they cannot remember specific events from their own pasts, their overall self-knowledge (for example, "I am generally comfortable in social settings") is reliable and can even be changed by new experiences. Moreover, both patients understand the concept of time: they know that there is a future and a past, they can tell time with an analog clock, and they know about their past and their future in a vague sense. What they lack is the ability to perform mental time travel.

Eleanor Maguire, Ph.D., and her colleagues at University College, London, have recently replicated what was observed with KC and DB and extended the observations in a more systematic fashion.[3] Five amnesic patients were tested for their ability to form mental images of novel experiences that might take place in the future in a familiar setting, such as a possible event in their lives over the next weekend. These patients were markedly impaired in their ability to do this—their mental images were vague and highly fragmented when compared to those reported by a control group of people of the same average age and level of education but without amnesia.

Another example can be found in people with severe depression. Researchers have known for some time that such individuals have difficulty in bringing to mind personal details from their past. Mark Williams, Ph.D., and his colleagues at the University of Wales showed in 1996 that people with clinical depression are also impaired in their ability to engage in episodic future thought, a discovery that may have important implications for understanding how prolonged depressive states are maintained. For example, an inability to envision a "brighter future" may contribute to sustaining depression.

Developing Mental Time Travel

A growing line of research has demonstrated that episodic memory emerges in children sometime between the ages of four and five. Around this age children display evidence of vividly recollecting details related to their memories, as opposed to simply remembering something in a more general sense. For example, a three-year-old is likely to have greater difficulty than a five-year-old in remembering which of two uncles gave him a treat the day before, even though both children would be quite confident that the event in question had occurred.

Recently, evidence has begun to accumulate suggesting that the ability to project oneself into the future emerges in concert with the ability to vividly recollect the past. Thomas Suddendorf, Ph.D., and his colleagues at the University of Queensland have shown that at about age five, children begin to be able to accurately report what they will or will not do in the future (for example, tomorrow), as well as what they have or have not done in the past (for example, yesterday).[4]

Seeking How Past and Future Are Related

These examples all suggest that the personal past and future are related in some way, but what is the nature of the relationship? Understanding this relation should provide important insights into brain disorders in which both past and future thought are absent and also into the functioning of the healthy mind in general. One possibility is that envisioning the future involves sampling the contents of our memories. In fact, Arnaud D'Argembeau, Ph.D., and his colleagues at the University of Liège have shown that people tend to envision past and future events in similar contexts.[5] For instance, if someone is asked to imagine a future event related to attending a lecture, he is likely to imagine himself sitting in a familiar classroom. What that classroom looked like, how big it was, and so on would play a large role in the imagined future scenario. One reason that amnesic patients, people with severe depression, and young children are unable to envision future events may be that they cannot rely upon memory for visual-spatial information when they attempt to construct images of the future.

In order to understand more about how this relationship between the past and the future operates, we recently conducted a functional neuroimaging study at Washington University in St. Louis.[6] We were specifically interested in whether thinking about the future and remembering past episodes involved similar regions of the brain.

Functional neuroimaging techniques, such as positron emission tomography (PET) and functional magnetic resonance imaging (fMRI), allow neuroscientists to examine brain activity associated with mental activity. When participants in a research study engage in a given cognitive task, PET or fMRI can provide information about the level of cerebral blood flow (PET) or blood oxygenation level (fMRI) in the particular parts of the brain involved in performing the task. These techniques measure aspects of metabolism, rather than directly measuring the activity of neurons, but the information they provide allows researchers to infer which brain regions are involved in a given task.

In the typical design of a neuroimaging study, brain activity associated with two tasks is contrasted with the hope of isolating the brain regions that are important for the cognitive process of interest. In most cases, researchers attempt to contrast a pair of tasks that are similar to one another but vary in one key way. For instance, in order to identify the brain regions that are important for episodic memory, many researchers choose to contrast an episodic memory task with one that does not involve a specific personal episode. An episodic memory task might require a person to recount his experiences on the day of his college graduation. The comparison task might involve simply stating the name of the college he attended. Both tasks require a person to retrieve a personal memory, but naming the college does not involve recollecting experiences at a specific time and place in the past.

Our goal was to identify brain regions that might be important for representing oneself in time and then to examine those regions to see whether or not they are similarly engaged by past and future thought. To accomplish this goal, we asked study participants to perform a set of three tasks while lying in an fMRI scanner. In two of these tasks, participants viewed a series of event cues (for example, birthday party) and were asked to envision either a personal memory of that kind of event or one that might take place in the future. Brain activity common to both tasks (past and future) was contrasted with that observed during a third task. This task involved similar processes, such as mental construction of lifelike scenarios, but did not involve representing oneself in time. Our specific control task required participants to imagine former U.S. president Bill Clinton. Clinton was chosen because pretesting showed that he is easy to visualize in a variety of situations.

The Brain During Past and Future Thought

As a result of our study, we learned that several regions in the brain's posterior cortex were similarly engaged during personal past and future thought, but not during the control task. As shown in Figure 1 (on page 24), these regions were located in the occipital cortex (A), the posterior cingulate cortex (B), and the medial temporal cortex (C). We found this very interesting, because previous research had shown that these same regions are consistently engaged during such tasks as autobiographical memory and mental navigation of familiar routes. We hypothesize that asking study participants to envision a personal future scenario likely requires similar processes. That is, in order to effectively generate a plausible image of the future, participants reactivate images (such as a familiar lecture hall) stored in the posterior cortical regions. Questionnaires that study participants filled out after the experiment corroborated this hypothesis: participants tended to imagine future scenarios in the context of familiar settings and familiar people.

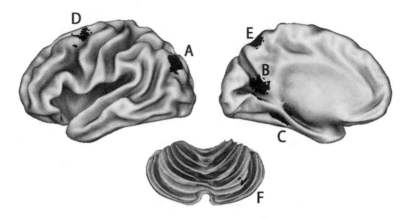

Figure 1 Brain regions showing similarities and differences during episodic future thought and remembering. Regions showing similarities appear within the superior occipital cortex (A), posterior cingulate cortex (B), and medial temporal lobes (C). Regions showing differences appear within the lateral premotor cortex (D), medial posterior precuneus (E), and right posterior cerebellum (F).

Credit: Courtesy of Karl K. Szpunar and Kathleen B. McDermott

Hypotheses based on exploratory research, such as identifying the neural relation of past and future thought, are significantly strengthened if the results can be replicated. Indeed, Donna Rose Addis, Ph.D., and Daniel Schacter, Ph.D., at Harvard University recently presented a similar set of data.[7] In their study, they demonstrated that mental construction of specific past and future episodes led to equivalent activity within brain regions similar to those that we identified.

One salient difference between past and future thought is that recollecting a personal memory requires reconstructing, as a whole, an event that has already taken place, but imagining a future event requires actively and continuously constructing a new scenario. The event has not yet taken place, so it is up to the individual to decide on a moment-to-moment basis where the event is taking place, who is there, and what they are doing. This difference was also demonstrated in our research.

As can be seen in Figure 1, we learned that several brain regions were more engaged when participants envisioned the future than when they recollected the past, although in both cases they were more involved than during the control task. These regions were found within the lateral premotor cortex (D), the medial posterior precuncus (E), and the posterior cerebellum (F). Previous research has suggested that each of these regions plays an important role in imagining a sequence of behavioral actions (for example, imagining tapping one's fingers or retracing one's jogging route). In short, we found that the act of mentally creating a future behavioral sequence—on the fly—appears to engage such brain regions to a greater extent than simply reflecting on a sequence of actions that have already taken place. Addis and Schacter also reported similar patterns of activity.

We were interested to find that not a single brain region was more engaged while recollecting the past than during thinking about the future. Addis and Schacter, whose research corroborated this, reported that the hippocampus (a structure important for pulling together the contents of memory to form a coherent mental image of the past) was more engaged during future thought than during recollection, and Maguire and her colleagues have also suggested that the hippocampus might be important in imagining future scenarios.

Insights from Neuroimaging

Advanced neuroimaging techniques can provide new insights into long-standing questions about cognition. Researchers who study memory have long known that a variety of factors influence how well we are able to learn (that is, encode) new information. For instance, a person who is asked to tie a new piece of information to knowledge about herself will generally remember that information better than someone who does not engage in such self-referential processing. Participants in a memory experiment who rate a list of adjectives (for example, "old") for self-descriptiveness will remember the list better than participants who think about the adjectives in a purely objective manner (for example, deciding whether "old" means the same as "aged"). Until recently, this effect on encoding has only been inferred from memory tests showing that information processed during learning in a self-relevant manner is remembered better during a later test than information that did not have this added "boost." Neuroimaging techniques now allow researchers to examine the influence of such variables during the actual learning phase. By observing brain activity associated with learning, we can find out whether there are any unique neural signals that may enhance encoding when people relate incoming information to themselves. This will be an exciting area for future research and has implications for education and learning of all kinds.

We have seen how brain regions known to be important for rein-stating past experiences appear to play an important role in constructing future personal episodes. This discovery confirms what makes intui-tive sense: thinking about the personal past and thinking about the personal future are closely related activities and involve similar processes. Of course, the story is never completely black and white, because we have also observed brain regions that show differences in neural activity as study participants think about the past and the future. Converging results from many studies will surely enhance understanding of how our brains discriminate memories of the past and imagination of the future.

Looking to the Future

Beginning with the pioneering work of Hermann Ebbinghaus,[8] a late-nineteenth-century German psychologist, students of psychology and neuroscience have expended more than 100 years of thought and careful experimentation toward an understanding of human memory. However, they have made surprisingly little inquiry into our ability to mentally represent the future. The coming years will surely see an increase in the number of studies of how the brain enables us to mentally repre-sent future thought and how future thought may be related to memory. Interesting questions and answers will likely arise from a wide range of research disciplines.

For example, one of the most common complaints associated with aging is memory loss. But do the elderly experience similar difficulties in thinking about the future? The answer to this question could have important implications for programs designed to improve older adults' health related behavior. For instance, older adults are better able to remember to take their medication if they have previously imagined performing the task. This implies that they can imagine the future, but how well? If older adults do have deficits in thinking about the future, would it be possible somehow to enhance their ability to do so?

As another example, think about your friends and colleagues. Is there someone who stands out as having a particularly good memory? What about someone with poor memory? People differ greatly in their capacity for remembering. What about future-directed thought? Are some people better able to imagine specific instances in their future and, consequently, better able to focus their behavior in order to accomplish their goals? On the other hand, might deficits in the ability to vividly envision the future have a negative impact on motivation?

We have had many enthusiastic discussions with our colleagues about directions for future research. In one of these, Washington University researcher Jason Chan suggested that understanding the relation between memory and future thought may provide insight into the creative capacity of the human mind. Authors of science fiction novels, for instance, are able to envision extremely detailed future scenarios. Might such an ability to vividly imagine the future be related to a corresponding strength in memory? Research on memory and on creativity are both part of studying human cognition, but until now little crosstalk between these two disciplines has taken place.

We believe that this is an exciting new approach to the study of the human mind and one that will greatly benefit from behavioral experimentation, clinical and neuropsychological observation, and recent advances in neuroimaging techniques.

CHAPTER 3

Prying into Prions

Challenges, Gloom, and Hope for Treating Deadly Diseases

by Scott P. Edwards

Scott P. Edwards is a health and medical writer based in Holliston, Massachusetts. He has written for *BrainWork* from Dana Press, as well as for Harvard Medical School, Brigham and Women's Hospital, the Mayo Clinic, the Robert Wood Johnson Foundation, and the H. Lee Moffitt Cancer Center. He can be reached at scottpedwards@verizon.net.

The rare but insidious prion diseases, so named for the tiny misfolded protein molecules believed to cause them, are invariably fatal. Although not one effective treatment for these diseases exists, new research is beginning to explain how the mutant proteins destroy brain tissue. The author explores this research, setting out the obstacles causing a cloud of pessimism to hang over many scientists and the successes that for some researchers offer the first glimmer of hope.

AMONG THE MOST MYSTERIOUS and virulent diseases known, prion diseases turn the brains of otherwise normal, healthy people into a spongy mush—and, without fail, kill them. While mad cow disease, Creutzfeldt-Jakob disease, fatal familial insomnia, and other prion diseases are exceedingly rare, scientists have spent the past twenty-five years searching for answers to how an enigmatic protein could cause such destruction. Researchers are now beginning to understand the role that mutant prion proteins play in these brain-wasting diseases, and though the road to a cure is full of obstacles, a bit of the gloom is starting to lift.

Chain Reaction of Misfolded Proteins

The idea that a protein could be the cause of an infectious disease was revolutionary until the early 1980s, when Stanley Prusiner, M.D., of the University of California–San Francisco School of Medicine introduced the concept of prions, for which he won the Nobel Prize in 1997. Prions, says Prusiner, are "tiny protein molecules that seem to cause a variety of slow-acting and inevitably fatal diseases in animals and humans." Prusiner named these proteins "prions," short for "proteinaceous infectious particles."

Found in the brains of all living animals, prions contain no DNA or RNA, the building blocks of the genetic code. In their normal form, prions are natural components of the body, although their function is still largely unknown. But when they become twisted or folded in large numbers, over time, they lead to diseases that are characterized by holes in the brain where neurons have died and by rapid loss of mental abilities.

Every protein in the body has a specific conformation, or shape, and many of them contain amino acids that allow them to change the shape of other proteins. Prions exist in two distinct conformations in the brain: a "wild type" that everyone possesses and a densely folded, infectious type. The infectious form of the protein recruits wild-type proteins to misfold, causing them to be infectious as well. Once this occurs, a massive chain reaction takes place in which increasing numbers of prions in the brain misfold and become infectious, triggering the cascade of deadly neurodegenerative effects.

In about 85 percent of human prion disease, the misfolding of proteins, called cellular prion proteins (PrPc), into a disease form, or isoform (PrPSc), occurs sporadically. Another 14 percent of human disease occurs from familial forms, associated with mutations of the gene that encodes prion protein. The remaining 1 percent of human prion disease is transmitted by eating food made from an infected animal, such as a cow. Whether sporadic, inherited, or transmitted, prion diseases all induce the buildup in the brain of toxic plaques that cause the widespread death of neurons.

In 2003, Stanford University geneticist Gregory S. Barsh, M.D., discovered that abnormal prion proteins, unlike their normal wild-type counterparts, are folded in such a way that enzyme target sites are hidden. In wild-type prions, enzymes degrade the proteins at the end of their lives. The folds in the abnormal prions, however, make them resistant to enzyme destruction.

Certain other neurodegenerative diseases feature a toxic buildup of other types of protein—notably Alzheimer's, in which the protein beta-amyloid forms plaques. Some researchers are investigating the idea that beta-amyloid might have a folding procedure similar to that of the abnormal prion.

The best-known prion disease is mad cow disease, formally known as bovine spongiform encephalopathy (BSE). Other animal prion diseases include scrapie, which affects sheep and goats, and chronic wasting disease, which afflicts deer and Rocky Mountain elks. Even rarer forms of the disease affect minks, monkeys, lemurs, household cats, and feline species in zoos. Among human prion diseases the most common are Creutzfeldt-Jakob disease, new variant Creutzfeldt-Jakob disease, fatal familial insomnia, and Gerstmann-Straussler-Scheinker disease. (To learn more about prion diseases, read "Prion Diseases: Rare but Deadly" on page 37.)

New Developments Show Promise

While not one effective treatment for prion diseases exists (the only available treatments aim to ameliorate symptoms), several recent research studies show promise. Among them are prion depletion, therapeutic vaccines, and inhibition of the protein fragmentation process.

In February 2007, researchers led by Giovanna Mallucci, Ph.D., of London's Institute of Neurology, reported a study showing that depleting prions in laboratory animals early in the disease process, before neuronal loss is widespread, may be an effective therapy. In mouse models, Mallucci and her colleagues found that an enzyme called Cre recombinase, which is used to modify genes and chromosomes, selectively inhibited the gene for PrPc so that no more of the protein is produced. Research efforts are under way to develop a drug that can safely induce this process.

"By depleting prions in the brain," says Mallucci, "the process of conversion [from normal to abnormal prion protein] is blocked. Depleting PrPc removes the substrate for ongoing replication and prevents continued production of the toxicity."

Mallucci says that for this approach to have any positive effect, prions would need to be depleted in mice up until about 60 percent of the way through the incubation period—before neuronal loss is established and when predegenerative changes can be reversed. Her study, published in the February issue of *Neuron*, showed that cognitive and behavioral impairments, as well as neurological pathology, could be reversed in mice if prion depletion is done early enough in the disease process. While scientists say the physiological role of the normal prion protein is unclear, Mallucci's team found that PrPc depletion had no effect on behavior or cognition in uninfected mice.

"Our findings of early reversible neurophysiological and cognitive deficits occurring prior to neuronal loss open new avenues in the prion field. To date, prion infection in mice has conventionally been diagnosed when motor deficits reflect advanced neurodegeneration. Now, the identification of early dysfunction helps direct the study of mechanisms of neurotoxicity and therapies to early stages of disease, when rescue is still possible," Mallucci and her colleagues wrote in their paper in *Neuron*.

In addition, she says that her research may also lead to preclinical testing of therapeutic strategies that could be helpful for early intervention in humans with prion diseases, particularly variant Creutzfeldt-Jakob disease. This could not only halt clinical progression of the disease but also reverse cognitive abnormalities, she adds.

Howard Federoff, M.D., Ph.D., a prion researcher at the University of Rochester Medical Center, recently reported that a therapeutic vaccine can slow the progression of prion diseases in mice. Federoff and his colleagues engineered a virus that carries the genetic code for producing antibodies that bind to and attack prion proteins. They injected this virus into the brains of mice, which allowed immune cells in their brains to produce antibodies based on the code. They then injected infectious mouse prion proteins into the bellies of these mice, which traveled to their brains and caused other normal proteins to misfold and form toxic plaques.

Both the test mice and control mice that did not receive the therapeutic vaccine died. The vaccinated mice, however, survived about 30 percent longer (260 days compared to 200 days for the control mice). Says Federoff, "We delayed the emergence of disabling clinical signs and substantially delayed death." The antibodies produced by the engineered virus likely work by binding to infectious prions and preventing the formation of the toxic plaques. According to Federoff, this approach might be effective for people who have variant Creutzfeldt-Jakob disease.

While conventional thinking holds that prion diseases worsen by converting ever more good proteins to misfolded bad forms, Tricia

Serio, Ph.D., and her colleagues at Brown University say that fragmentation of prion complexes is another crucial step in the disease process. In the February 2007 issue of *PLoS Biology,* the peer-reviewed, public-access journal of the Public Library of Science, Serio writes that a single protein plays a major role in deadly prion diseases by smashing up clusters of infectious proteins, creating the "seeds" of destruction.

According to Serio, Hsp104, a molecule required for prion replication, may play the role of "protein crusher." "To understand how fragmentation speeds the spread of prions, think of a dandelion," Serio said. "A dandelion head is a cluster of flowers that each carries a seed. When the flower dries up and the wind blows, the seeds disperse. Prion protein works the same way. Hsp104 acts like the wind, blowing apart the flower and spreading the seeds." Her team found in laboratory studies that Hsp104 crushes complexes of a yeast protein called Sup35 that is similar to human prion protein. Prions still multiply without this crush-induced fragmentation, she cautions, but they do so at a much slower rate. If a drug could safely inhibit this fragmentation process, it could substantially slow the spread of the infectious prions that cause Creutzfeldt-Jakob disease and other human prion diseases.

A paper in the February 2007 online edition of the *Proceedings of the National Academy of Sciences* offers further insight into how normal prions are converted and produce amyloid plaques, possibly leading to means to intervene in this process. Studying this same Sup35 protein that Serio's team used, researchers at the Scripps Research Institute, led by Ashok Deniz, Ph.D., found that the protein in its native state lacks a specific structure and forms intermediate shapes during the conversion of normal proteins to amyloid plaques found in Creutzfeldt-Jakob disease and variant Creutzfeldt-Jakob disease, as well as Alzheimer's, which is not a prion disease.

Deniz says this intermediate stage of the process is critically important, since "no single native unfolded protein is capable of initiating the amyloid cascade [by itself] because of this constant shape-shifting." This knowledge, Deniz says, may help scientists in the search for potential new therapeutic agents that intervene during this shape-shifting phase.

Challenges, Gloom, and "Reasonable Expectations"

Despite such imaginative and innovative efforts, human therapies are still a distant possibility, and gloom persists. In the years since Prusiner coined the term "prion" in the early 1980s, scientists have indeed learned a great deal about prion diseases. We now know that prions contain no nucleic acid (in contrast with most infectious diseases, which are spread through either a virus or bacteria, which contain DNA and RNA information) and that infectivity is associated with PrPSc. That said, however, numerous challenges face the scientific community in developing effective, appropriate treatments for prion diseases. Chief among them:

- *Prion structure.* Scientists know the 3-D structure of the normal prion protein, but not the structure of its disease form (PrPSc). If they can find out more about the PrPSc structure, they may be able to design effective drugs to counteract it, says Richard Johnson, M.D., of the Johns Hopkins University School of Medicine, who served as a consulting neurologist on a Public Health Service epidemiological study of human prion diseases and has testified before Congress about them.

- *Drug delivery.* Because prion diseases manifest themselves in the brain, drugs to treat them must be able to cross the blood-brain barrier. As scientists develop potential new drugs, they must find ways to deliver them to specific brain areas, and at high enough levels to prevent disease or to clear diseased proteins.

- *Misfolding.* Scientists know that the abnormal prion proteins form plaques in living animals and that they behave differently than normal proteins do, but they do not yet understand why the prion protein misfolds in the first place.

- *Diagnosis.* Most cases of prion diseases are not diagnosed in their early stages. Because they have long incubation periods—some up to fifty years—the buildup of abnormal prions in the brain has a cumulative effect. Thus, by the time most patients are diagnosed, dementia and other irreversible damage have set in.

- *Rapid progression.* Once symptoms appear, the disease progresses rapidly. From week to week, says Johnson, patients lose their ability to work and remember names, among other symptoms. At this point it is too late for effective treatment. "Drugs at this stage," he adds, "would only preserve dementia, which is not what we're looking for."

Federoff says that perhaps the most difficult challenge facing scientists is "developing reasonable expectations to be able to alter these diseases." Because of these challenges, a sense of pessimism pervades the prion research community. "Think about a misfolded protein," says Johnson. "What kind of drug could unfold it or make it not fold in the first place? It's hard to picture a drug that could do this."

Despite the sense of gloom among researchers, Federoff, for one, remains optimistic. Success, he says, depends on scientists' ability to more regularly diagnose prion diseases at their early stages, when there is greater potential for restoring or abating underlying problems in the brain. These therapeutic strategies, he adds, may be partially or fully effective.

"Take, for example, cancer therapies that extend life an additional three, six, twelve months," he says. "Any significant extension of functionality is an advance. We're not talking about cures, but extending life and functionality that will resonate with patients."

That is a goal worth working toward.

Prion Diseases: Rare but Deadly

Mad cow and other prion diseases—so named by Nobel Prize winner Stanley Prusiner, M.D., for the tiny protein molecules believed to cause these degenerative brain diseases—have been around for years. These diseases, although fatal, are also very rare and have caused what some scientists say is irrational fear.

Animal Prion Diseases

A number of common animal prion diseases have been identified. Mad cow disease, formally known as bovine spongiform encephalopathy (BSE), has a long incubation period (up to four or five years), but is usually fatal to cattle within weeks or months of onset of symptoms. BSE is characterized by spongelike changes in the brains and spinal cords of infected cattle that alter the animals' temperament, sometimes causing aggressiveness, swaying gait, and difficulty in rising. Scientists traced the cause of the disease to cattle feed prepared with diseased cow brains and spinal cord tissue. Though rare, BSE is transmissible to humans through the consumption of contaminated beef.

Scrapie (so-called because infected animals were observed scraping against fences) is a slow, progressive disease that causes central nervous system degeneration in sheep and goats. The first symptom of the disease is a slight change in animal behavior, followed by nervousness, aggressiveness, and separation from the flock. Animals with scrapie appear normal until they are startled, which causes them to go into convulsions. Over time, infected animals become uncoordinated, especially in their hind limbs. Once symptoms appear, the disease is generally fatal within six months. Scrapie is not believed to be transmissible from animals to humans.

Chronic wasting disease (CWD) is common in deer and Rocky Mountain elk in the tricorner area of Wyoming, Colorado, and Nebraska. The disease, first recognized in 1981, is characterized by weight loss, behavioral changes, excessive salivation, difficulty swallowing, excessive thirst, and frequent urination, as well as a lack of muscle coordination and head tremors in some animals. Most animals die within several months of onset. No cases of human prion disease have been associated with CWD.

Other forms of animal prion diseases include transmissible mink encephalopathy, feline spongiform encephalopathy (a neurological disease in household cats and feline species in zoos that has been linked to BSE), and a form of the disease in nonhuman primates such as macaque monkeys and lemurs. Most of these diseases are extremely rare.

"Insidious" Human Diseases

In humans, the most common forms of prion diseases are Creutzfeldt-Jakob disease (CJD), variant CJD, fatal familial insomnia, Gerstmann-Straussler-Scheinker disease, and kuru.

Like animal prion diseases, CJD is characterized by spongelike holes in the infected brain. The disease is rare, affecting one person per million worldwide, with about 250 to 300 new cases in the United States each year. CJD is hard to diagnose because in its early stages the symptoms are similar to those of other neurodegenerative diseases such as Alzheimer's, Huntington's, and Parkinson's diseases. Early symptoms of CJD include mood swings, depression, memory difficulties, social withdrawal, and anxiety, as well as visual disturbances, fatigue, and bizarre behavior. As the disease progresses, CJD patients suffer from dementia, muscle paralysis, slurred speech, difficulty swallowing, and eventual blindness. Some patients fall into a coma. All die. Most cases of CJD are considered "sporadic" because the disease occurs for no known reason.

Variant CJD (vCJD) is a relatively new disease, first discovered in 1996. This form of CJD, the human form of mad cow disease, typically strikes younger people, with an average age of onset of twenty-nine years (compared to sixty-five for other forms of CJD). Early in their illness, people with vCJD experience psychiatric symptoms, including depression and schizophrenic-like psychosis. Nearly half of the people diagnosed with vCJD have reported unusual sensory symptoms like "stickiness" of the skin. The disease is also characterized by unsteadiness, difficulty in walking, and involuntary movements. These symptoms develop as the illness progresses, and toward the end stage of the disease many people are completely immobile and unable to speak. Few cases of vCJD, which is caused by an inherited mutation of the prion protein gene, have been reported in the United States.

Fatal familial insomnia is a rare genetic disorder that affects the thalamus, the region of the brain that controls sleep and wakefulness. Symptoms are directly related to a malfunction of the thalamus's sleep mechanism. Over the first few months, patients suffer from psychiatric problems, including panic attacks and bizarre phobias. The second stage includes hallucinations, panic, and agitation. The third stage is characterized by near-total insomnia and significant weight loss. During the final stage of the disease, which lasts about six months, patients suffer from dementia, total insomnia, and sudden death. The time from onset of symptoms to death is about eighteen months.

An extremely rare disease found in only a few families around the world, Gerstmann-Straussler-Scheinker disease (GSS) is almost always inherited. Early symptoms include lack of muscle coordination, or ataxia, and difficulty in walking. The disease progresses to more pronounced ataxia and dementia. Other symptoms include speech difficulties, rigid muscle tone, and vision and hearing loss. Some people with GSS also have Parkinson's-like features. The disease causes extreme disability and eventual death, often after a patient develops a secondary infection or slips into a coma.

Kuru has now largely been eradicated, but epidemic levels occurred in the 1950s and 1960s among indigenous peoples of the New Guinea highlands because of ritualistic cannibalism practices in which the natives consumed tissues, including the brains, of their deceased relatives. The disease affected the brain's cerebellum, which is responsible for muscle coordination. People infected with kuru were unable to stand or eat, and they died in a comatose state.

While all of these human prion diseases are rare, Johns Hopkins University neurologist Richard Johnson, M.D., says they are particularly insidious. "You get them, you die," he says. "There's no way around it."

CHAPTER 4

Protecting the Brain from a Glutamate Storm

by Vivian Teichberg, Ph.D., and Luba Vikhanski

Vivian Teichberg, Ph.D., a professor of neurobiology at the Weizmann Institute of Science in Rehovot, Israel, studies the physiological and pathological roles of glutamate in the brain. He was among the first to isolate and study a receptor protein to which glutamate binds in the brain, and to solve this receptor's primary structure. He is the recipient of several prizes, including the Somach Sachs Memorial Award and the French Academy of Sciences Professorship, and holds the Louis and Florence Katz-Cohen Professorial Chair of Neuropharmacology. He can be reached at vivian.teichberg@weizmann.ac.il.

Luba Vikhanski works as a science writer in the Publications and Media Relations Department of the Weizmann Institute. She is the author of *In Search of the Lost Cord: Solving the Mystery of Spinal Cord Regeneration* (Dana Press and Joseph Henry Press, 2001). Her news stories and feature articles have appeared in publications including *Scientific American*, *Nature Medicine*, the *New York Times*, and the *Jerusalem Post*. She can be reached at luba.vikhanski@weizmann.ac.il.

When a stroke or head injury releases a flood of the chemical messenger glutamate, the excess glutamate leaves damaged neurons in its wake. Israeli scientist Vivian Teichberg, Ph.D., has developed a new method that may protect the brain from this destruction by harnessing the brain's natural ability to keep glutamate levels in check.

THE HUMAN BRAIN is packed with a substance that needs to be treated like a handle-with-care explosive. Glutamate, one of the most abundant chemical messengers in the brain, plays a role in many vital brain functions, such as learning and memory, but it can inflict massive damage if it is accidentally spilled into brain tissue in large amounts.

Glutamate flow in the brain is normally kept in check by a system of damlike structures, which release a trickle of the substance only when and where it is needed. But burst a dam—as happens in stroke, head trauma, and some other neurological disorders—and the treacherous messenger floods the brain. The surge of glutamate radiates out from the area of original damage and kills neurons in nearby areas. The expanded damage can leave in its wake signs of impaired brain function, such as slurred speech and shaky movement.

Depending on the severity and location of the stroke or head trauma, recovery can be slow and incomplete. Now new hope is coming from a completely new approach to protecting the brain against the ravages of injury and disease. It consists of "mopping up" excess glutamate by boosting a natural process that the healthy brain already uses to safeguard itself from a glutamate overdose. If this concept is borne out in clinical trials, it might be helpful in treating a variety of acute and chronic brain insults and diseases.

Inside the Glutamate Storm

The amino acid glutamate is the major signaling chemical in nature.

All invertebrates (worms, insects, and the like) use glutamate for conveying messages from nerve to muscle. In mammals, glutamate is mainly present in the central nervous system, brain, and spinal cord, where it plays the role of a neuronal messenger, or neurotransmitter. In fact, almost all brain cells use glutamate to exchange messages. Moreover, glutamate can serve as a source of energy for the brain cells when their regular energy supplier, glucose, is lacking. However, when its levels rise too high in the spaces between cells—known as extracellular spaces—glutamate turns its coat to become a toxin that kills neurons.*

As befits a potentially hazardous substance, glutamate is kept safely sealed within the brain cells. A healthy neuron releases gluta-mate only when it needs to convey a message, then immediately sucks the messenger back inside. Glutamate concentration inside the cells is 10,000 times greater than outside them. If we follow the dam analogy, that would be equivalent to holding 10,000 cubic feet of glutamate behind the dam and letting only a trickle of 1 cubic foot flow freely outside. A clever pumping mechanism makes sure this trickle never gets out of hand: When a neuron senses the presence of too much glutamate in the vicinity—the extracellular space—it switches on special pumps on its membrane and siphons the maverick glutamate back in.

This protective pumping process works beautifully as long as gluta-mate levels stay within the normal range. But the levels can rise sharply if a damaged cell spills out its glutamate. In such a case, the pumps on the cellular membranes can no longer cope with the situation, and glutamate reveals its destructive powers. It doesn't kill the neuron directly. Rather, it overly excites the cell, causing it to open its pores excessively and let in large quantities of substances that are normally allowed to enter only in limited amounts.

One of these substances is sodium, which leads to cell swelling because its entry is accompanied by an inrush of water, needed to dilute

---

*The glutamate in MSG (monosodium glutamate), used in some foods, is related to brain glutamate but does not appear to get into the brains of adults very well. It can get into the brains of infants and be toxic to brain cells, however, which is why the FDA has not approved it for use in baby food.

the surplus sodium. The swelling squeezes the neighboring blood vessels, preventing normal blood flow and interrupting the supply of oxygen and glucose, which ultimately leads to cell death. Cell swelling, however, is reversible; the cells will shrink back once glutamate is removed from brain fluids. More dangerous than sodium is calcium, which is harmless under normal conditions but not when it rushes inside through excessively opened pores. An overload of calcium destroys the neuron's vital structures and eventually kills it.

Regardless of what killed it, the dead cell spills out its glutamate, all the vast quantities of it that were supposed to be held back by the dam. The spill overly excites more cells, and these die in turn, spilling yet more glutamate. The destructive process repeats itself over and over, engulfing brain areas until the protective pumping mechanism finally manages to stop the spread of glutamate.

That's precisely what happens in stroke or head trauma, each of which begins with a sudden injury to brain tissue that ensues when a blood vessel is ruptured or blocked by a blood clot. In trauma, the damage is inflicted by a blow to the head. If the damaged area, called the core, is small and not located in a vital region of the brain, it might not cause major harm. However, because the dead cells in the core spill out their glutamate, the core often becomes the center of a glutamate spill. While the center itself cannot be saved, the secondary damage triggered by glutamate release from damaged or dying brain cells could theoretically be prevented or at least limited, and perhaps even reversed. Medical management can help to prevent further damage, but effective neuroprotective drugs have, so far, been hard to come by.

Why Drugs Fail

Scientists may argue about the relative roles of glutamate and other chemicals, but research overwhelmingly suggests that excess release of glutamate is one of the earliest and most crucial steps in the destructive cascade of events in the traumatized brain and that glutamate's relationship to neurological damage is that of cause and effect. Furthermore, physicians have established that the chances for recovery from stroke or head trauma are better the lower the levels of glutamate are within the spaces between brain cells. These extracellular spaces are where brain cell axons (communication cables) send signals from one brain cell to another. Conversely, the higher the glutamate levels in these extracellular spaces, the poorer the outcome from stroke or head injury.

However, in focusing just on glutamate's destructive potential, research has overlooked the chemical's role as the brain's premier messenger, and that omission may be a key to why glutamate-blocking drugs, which prevent glutamate from overexciting brain cells, fail in clinical trials. Entirely blocking glutamate's activity in the brain after stroke or trauma may not be such a good idea: Along with stopping glutamate's demolition activity, the blocking shuts down all messages carried by glutamate, and these messages might be critical for the patient's recovery, as they appear to be essential for the initiation and maintenance of the brain's repair mechanisms.[1] In other words, it's possible that glutamate-blocking drugs have failed to promote recovery after stroke because they have been inadvertently blocking the brain's natural self-repair mechanisms.

Another reason for the failure of glutamate blockers might be timing. In rats, glutamate levels are excessively high for only about two hours after trauma, and a single injection of glutamate blockers during this time can promote recovery. In humans, in contrast to rats, the surge of glutamate may persist for hours or even days and therefore needs to be dealt with over a longer period of time. Yet in most clinical trials of stroke treatments, glutamate blockers were given just once. These blockers, as well as the other drugs, might have failed because even if they had a positive effect, they didn't take care of the glutamate spill, which continued to wreak its havoc after the drug had cleared from the brain.

Then we have the problem of delivering the drug to the area of brain damage. Drugs intended to protect the brain are purposely designed to enter the bloodstream and cross the blood-brain barrier, the impervious layer of cells that lines the walls of the brain's blood vessels and prevents most substances from entering our most well-guarded organ. However, in stroke and head trauma, the supply of blood to the damaged brain area is impaired because blood vessels are blocked, ruptured, or constricted. Therefore, even the most effective drugs are prevented from producing their beneficial effect because they cannot get to the site of injury.

Moreover, moving from rat research to clinical trials in humans presents an overall challenge. Unlike young and uniform laboratory animals, each human patient requiring a stroke or brain trauma medication is unique, and this variability makes it difficult to determine whether the drug is indeed effective. Strokes tend to occur when people are older and might have diabetes, heart disease, or other conditions that can interfere with a drug's activity in unpredictable ways. Other issues can arise in connection with head trauma, such as the presence of alcohol in the blood or brain of accident victims, which can also affect a drug's action.

Finally, developing a drug for stroke or brain trauma poses a major challenge because the drug must do significantly better than natural brain repair mechanisms, yet nature already does a reasonably good job of posttraumatic brain repair. So far, only the clot-dissolving drugs called tissue plasminogen activator (tPA) have met the United States Food and Drug Administration's requirements for both safety and effectiveness.

A New Approach

The disappointing performance of neuroprotective drugs for treating stroke in clinical trials was the starting point for recent research at the Weizmann Institute in Israel that has led to a radically new approach to battling excess glutamate. My (Vivian Teichberg's) research team has developed an innovative experimental therapy relying on the natural pumping mechanism that normally protects the brain against excess glutamate.

As mentioned earlier, much of this pumping removes glutamate from the extracellular spaces between brain cells by siphoning glutamate back into these cells—mainly the neurons—and their supporting cells, the glia. But an additional route for this protective mechanism has received much less attention from scientists. Glutamate pumps are also present on the outer surface of the brain's blood vessels, the surface facing the brain tissue. The fact that glutamate is being pumped from the brain into the blood circulation was demonstrated more than forty-five years ago. This was observed back in the 1960s by Soll Berl, M.D., from the New York State Psychiatric Institute, who found that if a small trace of radioactive glutamate is injected into the brain, it shows up outside the brain, in the peripheral blood, within one minute.

After we repeated Berl's experiment and confirmed his findings, our next goal was to harness this brain-to-blood pumping process for brain-protective purposes. Such a procedure has great therapeutic potential, because brain tissue is densely meshed with tiny blood vessels called capillaries. The human brain has more than 100 billion capillaries, with a total length of about 400 miles and a total surface area of 130 square feet—and a huge number of glutamate pumps! Still, toxic glutamate spills occur in stroke, head trauma, and many neurological diseases, which shows that this protective arsenal does not always rise to the task when things go badly wrong.

The question was how to increase the pumping of glutamate out of the brain and into the bloodstream to protect the brain. Shutting down a cellular pump is easy; making it work faster and better is much more tricky. As is true for many natural processes perfected by eons of evolution, these pumps already work very well, so enhancing such a process is no mean feat.

Our answer came from exploring the secrets of brain-to-blood pumping. It turned out that two efficient yet disarmingly simple mechanisms are in play. First, glutamate is pumped into the cells that make up the blood vessel wall. This step is performed by the glutamate pumps on the outer side of the brain's blood vessels, the side that comes into contact with the brain tissue. Because the blood vessel wall cells are very small in size, glutamate concentration inside them quickly builds up to very high levels, much higher than the chemical's concentration in the blood. That leads to the second step of the process, which stems from basic chemistry: Glutamate naturally flows by diffusion from areas of high concentration (in blood vessel wall cells) through the blood vessel wall into the circulating bloodstream, where the concentration is lower.

This insight led to the idea of accelerating the naturally occurring brain-to-blood pumping by lowering glutamate levels in the circulating blood. The hypothesis was that a larger difference in glutamate concentration would enhance the driving force for the chemical's removal, and more glutamate would flow out of the brain and into the circulating blood.

That was indeed what happened. We made use of the fact that certain blood enzymes can reduce blood glutamate levels by transforming glutamate into a different substance—a transformation that takes place in the presence of increased levels of certain compounds. When these compounds were injected into the bloodstream of rats, this transformation rapidly occurred and the animals' blood glutamate levels dropped by 50 percent—and so did glutamate levels in the brain's cerebrospinal fluid. Thus, the experiment strongly suggested that lowering glutamate levels in the circulating blood was an effective strategy for kicking the glutamate pumps' activity into high gear in order to "mop up" toxic glutamate spills in the brain.[2]

The next task was to test whether clearing excess glutamate from the brain by this method would protect the brain from glutamate's deleterious effects. In experiments in rats with traumatic brain injury, a natural compound called oxaloacetate, which reacts with the blood enzyme glutamate-oxaloacetate transaminase, was used to scavenge blood glutamate. Rats with brain injury treated by this method regained normal body function and recovered faster and more fully than rats with the same brain injury that received no medication to lower glutamate levels.[3]

We are currently developing noninvasive ways of assessing brain glutamate levels in humans that should help implement the new approach in clinical practice. Drilling a hole in a person's skull just to measure the glutamate is hardly an attractive prospect, but magnetic resonance spectroscopy, an analytical method that reveals the presence of various brain chemicals through the skull in a noninvasive manner, might offer a viable alternative.

Avoiding Problems

The proposed Weizmann Institute concept circumvents many of the problems encountered by previous brain-protecting drugs. Most of the trials of agents to modify excess glutamate were stopped because of side effects, primarily psychosis, but the natural compounds we tested are safe and cause no side effects. In addition, our method involves no concern about drugs crossing the blood-brain barrier or entering the brain despite impaired blood circulation. The added compounds act within the blood in order to protect the brain, rather than targeting the brain directly.

Nor does a concern exist about interfering with glutamate's beneficial action. Rather than blocking glutamate inside the brain, the method clears the excess chemical away from the brain into the blood, where it can no longer do harm. At the same time, the amount of glutamate needed for conveying messages of brain repair can remain in the brain.

No less important, this method aims to protect the brain by going to the root of the problem: preventing the glutamate storm. Whether it can be used alone, however, or in combination with other therapies, will depend on the timing. If it is applied immediately after stroke or injury, cleaning the brain of excess glutamate could be sufficient for preventing all subsequent problems, and no other drugs may be necessary. In many cases, however, the drugs cannot be given immediately. Strokes sometimes occur during sleep, and by the time the person wakes up and discovers the damage, precious time has been lost. Moreover, both strokes and head traumas can occur in areas where transportation to the hospital can take many hours. In such cases, a glutamate "mop-up" will still be relevant, but it may need to be accompanied by a "cocktail" of other drugs to counter the various types of damage already inflicted by the glutamate spill.

Toward Wider Applications

If proved effective, this new method may be helpful in treating not only stroke and head trauma but a variety of acute and chronic conditions—ones that involve the death of brain cells followed by a harmful glutamate glut. For example, a type of brain inflammation known as bacterial meningitis is successfully treated by antibiotics but often leads to neurological problems, such as hearing impairment and cognitive deficits, believed to be caused by excess glutamate. Toxicity from excess glutamate is thought to be a component of other conditions as diverse as hypoglycemia, some brain cancers, damage to a newborn's brain caused by interrupted oxygen supply during delivery, and exposure to nerve gas. Glutamate is probably also involved in chronic nerve damage in such conditions as glaucoma, amyotrophic lateral sclerosis (ALS), and HIV dementia. Though the mechanisms of excess glutamate release in these conditions are more obscure compared with acute brain insults, clearing surplus glutamate from the brain might be beneficial.

The method developed at the Weizmann Institute will soon be applied in clinical trials to be conducted by an Israeli biotechnical company. They will be designed to test whether the approach can protect the human brain against glutamate damage as well as it protects the brains of injured rats. Because the method works with nature and not against it, we hope that it will succeed where many others have failed.

CHAPTER 5

Cerebral Malaria, a Wily Foe

by Kayt Sukel

Kayt Sukel is a writer whose essays and articles have appeared in *Science*, *Memory and Cognition*, and *NeuroImage*, as well as the *Washington Post*, the *Christian Science Monitor*, and *National Geographic Traveler*. She can be reached at ksukel@hotmail.com.

Malaria is a preventable, treatable disease, but it's far from under control in many parts of the world, including Africa, where it is particularly lethal to children. Much of that toll is the result of malaria infiltrating the brain. But with increased attention and funding, research on malaria is flourishing in some unexpected areas, including understanding the interaction of the brain and the immune system.

HALIMA, A THREE-YEAR-OLD GIRL, was brought to the hospital in Kenya after running a fever for almost two days. At first, the fever seemed nothing to be particularly concerned about, so Halima's mother gave her paracetamol (acetaminophen) to bring down her temperature and left her in the care of an older sister while she went out to work on the farm. But when she returned a few hours later, she was unable to wake her child. She shook her gently and Halima's eyes opened, but the girl stared blankly ahead, unable to make eye contact. Her sister told their mother that Halima had had a convulsion earlier, her arms and legs jerking uncontrollably for several minutes before her body went limp. It was then that the mother began the arduous four-hour trek to the hospital for treatment.

The hospital physician noted that Halima's fever was over 103° Fahrenheit and her gaze was blank and roving. Shortly after the initial examination, Halima began convulsing again, and the physician administered an anticonvulsant drug. The physician listened to her mother's story—one that physicians hear day after day in Kenya, Malawi, and other parts of sub-Saharan Africa—and made a tentative diagnosis of cerebral malaria, a form of severe malarial infection. He began presumptive treatment with quinine and intravenous fluids and waited to see if Halima would be one of the lucky ones who manage to survive.

Hundreds of millions of people contract malaria each year, primarily in the poor countries of sub-Saharan Africa. Most are sick for only a few days. But in a small percentage of those infected, including Halima, the malarial parasites will attach to blood vessels and capillaries in the brain, causing coma, neurological damage, organ failure, and, often, death.

How does malaria produce such profound symptoms? Could the body itself be causing damage in its attempts to keep the parasite at bay? What lies behind why some people are stricken with one of the most brutal variations of the disease, while others are not? Scientists are currently working to answer those questions by studying the malaria parasite itself, the human immune response to this intruder, and the variations in how people experience the disease. Promising research on cerebral malaria is taking place around the globe, in university laboratories from the United States to Australia and in the field in Africa.

A Terrible Prognosis

Malaria is an infectious disease caused by the release of protozoan parasites into the bloodstream by the bite of a parasite-carrying *Anopheles* mosquito. After an incubation period of one to four weeks, initial malaria symptoms begin that usually include fever, headaches, vomiting, chills, and general malaise, similar to the flu. These symptoms are caused by the release of the parasites' products into the bloodstream. Most people, if treated, recover relatively easily, but the unlucky others, like Halima, will develop the disease's more severe form, cerebral malaria, in which the parasite-infected red blood cells attach in large numbers to the circulatory vessels of the brain.

What is the prognosis for a child whose malarial infection has localized in the blood vessels of the brain? If not immediately treated, cerebral malaria is likely to be fatal. But even with treatment, the physician can only wait to see the outcome. For most children, the coma will reverse and they will recover. But within three to seven years, approximately a quarter of those who do recover will show impairment in memory, attention, and other cognitive skills.

Halima became more responsive after fifteen hours of the simple treatment with quinine and fluids. By the end of two days, her alertness had continued to improve, but she was still unable to fix her gaze or follow a moving object. She also experienced weakness on her right side. A month after discharge from the hospital, her vision had improved, but she still walked with a limp. And one year later, she was developing on course with her peers but had developed epilepsy, a disorder that will require medication for the rest of her life to control what would otherwise be unprovoked, aggressive seizures.

Sadly, Halima's recovery is considered a relatively good outcome, because a significant portion of those who contract cerebral malaria each year—an estimated 15–20 percent—die of the disease.

Developing Immunity

The form of malaria that invades the brain is usually caused by *Plasmodium falciparum*, one of the four malarial parasites that can infect humans. According to Charles Newton, M.D., a physician and researcher with the KEMRI–Wellcome Trust Research Programme in Kilifi, Kenya, more than 2 billion people are exposed to falciparum malaria in the world annually. The more than 500 million episodes of the disease each year—some of them repeated illnesses in the same person—result in at least 1 million deaths, making falciparum malaria the most fatal parasitic disease in the world. Young children in sub-Saharan Africa bear the brunt of this burden.

Over time, most people native to endemic malarial areas, who become infected with the uncomplicated form of the disease over and over again, build up a resistance to it because their bodies develop an effective antibody defense. Although this immunity has been well established epidemiologically, scientists do not yet fully understand the mechanisms behind it.

"Malaria doesn't induce a 'sterilizing immunity,' like measles or smallpox, where you get it once and never get it again," says David Sullivan, M.D., an associate professor at the Johns Hopkins Bloomberg School of Public Health. He explains that as the body develops antibodies that fight malarial infection, the parasites lose the ability to attach to the lining of the blood vessels in numbers as great as in severe falciparum disease.

But those antibodies do not totally prevent a person from contracting malaria again. Instead, they keep the number of infected blood cells below the threshold that would cause the more severe forms of the disease. Children under the age of six years and people who are not native to areas where malaria is pervasive have not yet fostered this immunity by building up enough antibodies, and so they are left susceptible to the more severe manifestations of the disease, such as cerebral malaria.

Questions for Research

To date, efforts to prevent or at least control malaria primarily involve spraying with insecticides, particularly spraying bed nets, and administering multiple doses of antimalarial drugs such as Malarone and Lariam to people traveling to areas where malaria parasites are rampant. "These methods are ways of preventing malaria," says Terrie E. Taylor, D.O., a Michigan State University Distinguished Professor who for more than twenty years has spent the African rainy season, January to June, treating patients at the Queen Elizabeth Central Hospital in Blantyre, Malawi. "But the caveat is that they have to be sustainable for an indefinite period."

Malaria's incidence has soared over the past two decades as a result of mosquito resistance to pesticides and parasite resistance to common antimalarial drugs. Tied to this increased rate of regular malarial infection is the growing number of people contracting, and dying from, cerebral malaria. This has made the development of new methods and treatments to help stanch the disease's substantial mortality and morbidity rates a focal project not only in areas where malaria is endemic but for several deep-pocketed pharmaceutical firms and charitable organizations across the world.

What do we know of the malarial parasite that brings about the cerebral form of the disease? Surprisingly little. "Malaria is a very complicated disease with a complicated life cycle," says Diane Griffin, M.D., Ph.D., the Alfred and Jill Sommer Chair of Microbiology at the Johns

Hopkins Bloomberg School of Public Health and director of the Malaria Research Institute. "We need to understand all components of that life cycle better in order to control it."

As more attention and funds have been allocated to fight the spread of malaria in general, and its crueler manifestations such as cerebral malaria, increased understanding of the mechanisms underlying how the malaria parasite can attack brain function is raising new hopes for preventing malaria's most fatal forms. But surprisingly, greater knowledge about what scientists do not know about malaria is also helping to drive research in the right direction, both in clinical research and in investigations into the human immune response to malarial parasites.

Why Halima, as opposed to one of her friends, parents, or siblings, would come down with cerebral malaria is only one of the many questions about how cerebral malaria operates. At a more basic level, researchers are looking to solve how a parasite that remains in the blood vessels in the brain but does not invade the brain tissue itself can bring about such profound symptoms and neurological problems that may include, besides those Halima experienced, aphasia, ataxia, cortical blindness, coma, and, finally, death. When answers are found, scientists can develop better methods to prevent the more deadly types of malarial disease.

Surprises, and a Better Diagnosis

Terrie Taylor focuses on clinical research with children who have cerebral malaria in Blantyre, an industrial center in Malawi. To better understand how malaria can cause coma and death, she and her team have embarked on an autopsy study of children, with the hope of identifying how cerebral malaria damages the brain. She and her colleagues have not yet answered many of the questions they had when they started the study, but they have made interesting discoveries that will change how cerebral malaria is diagnosed and examined.

One area that Taylor's team has focused on is the endothelial cells, the unique, flat cells that form the lining of blood vessels. These cells are an integral part of the blood-brain barrier, which protects vulnerable brain tissue from invaders. "We know that the small blood vessels of the brain have parasitized red cells that adhere to the endothelial cells in those vessels," says Taylor. "But in terms of what is causing what, lots of different facets need to be considered."

Taylor thinks that several problems could be responsible for the more complicated forms of malaria and their neurological symptoms, for example a lack of blood flow to the brain or slower blood flow resulting in brain damage, swelling, and inflammation of clogged blood vessels, or perhaps damage stemming from seizures. She hopes to learn more as the autopsy study continues.

One of the most significant discoveries from Taylor's study is that approximately one-quarter of children autopsied, who met the standard case definition for cerebral malaria before they died, actually died of completely unrelated infections or diseases. "This really calls into question a lot of work that's been done on severe malaria to date," says Taylor. "The studies might have included patients who were not suffering from malaria at all because the researchers were using case definitions that lacked precision."

This valuable insight led other physicians, partnered with Taylor, to determine that the only clinical difference between those children who died of cerebral malaria and those with malarial infection who died from other causes was a complication of cerebral malaria that damages the retina, called malarial retinopathy. Using just an ophthalmoscope, clinicians can check the eyes of sick children to see if telltale whitening of the eye's infected blood vessels and swelling of the optic nerve are apparent. If so, the child is most likely suffering from cerebral malaria, not from another disease with just an incidental malarial infection.

"It's important to note that not all malarial infection is equal to malarial disease," says Taylor. "There are plenty of people walking around with malaria parasites that are incidental and have no relationship to the disease they might have." Improving the case definition by checking for malarial retinopathy will help clinicians correctly diagnose cerebral malaria and administer treatment more quickly and effectively, and it will also enable scientists who are studying malaria to be sure they are including in their studies only those who truly have the disease, thereby removing potential ambiguity from their research.

Taylor's research has yielded other interesting results as well. She found that those children in the autopsy study who did die of true cerebral malaria all had parasites attached to the brain's blood vessels. But she identified two distinct forms of this pathology, the first showing just the parasites sequestered in the blood vessels and the second showing the parasite-filled vessels along with hemorrhages, clots, and other tissue damage. "We have a long way to go until we can work out the cause-and-effect relationship," says Taylor. "We are not certain if these are two separate disease processes or perhaps different stages of the same process."

Understanding the Deadly Parasite

While Taylor's group has focused on clinical research, David Sullivan and Monique Stins and their research team at Johns Hopkins are examining the biological mechanisms of the *Plasmodium falciparum* parasite and how it can cause coma and other neurological problems even though it is not able to cross the blood-brain barrier. According to Sullivan, "The question is, what are the signals that the malaria parasite is able to send through the endothelium so that it does not invade the brain tissue itself but is still able to choke off the brain's function?"

Their laboratory has found that malaria parasites, after sticking to the receptors on endothelial cells and plugging thin blood vessels called "capillaries," also release proteins and molecules that make the blood vessel wall leaky. "In some ways, malaria is a disease at the capillary level," says Sullivan. "That's how the parasite homes in on the brain, blocking the blood vessels, and causes the more severe cerebral malaria."

More than 100 different types of highly variable *Plasmodium falciparum* proteins have been identified that cause sticking to vessels. In the future, Sullivan hopes to understand how malarial parasites are able to send messages from the endothelial cells to the microglial cells and neurons in the brain, causing coma and other neurological consequences of severe malarial infection. "We're still looking for the messenger molecules that turn off the brain in such a reversible fashion, without a lot of inflammation," he says. "Even when the patient is in a coma, the brain tissue itself appears to be normal."

The Link to Neurological Impairment

But the influence of cerebral malaria does not end if and when the child recovers. The disease can cause brain damage that will stay with those afflicted for the rest of their lives. Charles Newton, who both sees patients and does research full-time in Kenya, studies the link between cerebral malaria infection and later neurological impairment. His group has found that nearly one-quarter of children who survive cerebral malaria will go on to show problems with memory, attention, and other cognitive function years later. Scores of children like Halima will live out their lives with the aftereffects of the disease, aftereffects that hinder their ability to finish school, find and keep employment, and take care of their own families.

Newton's group has also shown an association between severe malaria and the development of epileptic conditions like that with which Halima was diagnosed a year after her initial recovery from cerebral malaria. At first glance, lack of blood flow (ischemia) due to blocked blood vessels may seem like the obvious cause for this kind of neurological damage, but Newton argues that other features of the disease are at play. "Although reduction of blood flow due to blockage of blood vessels may contribute, it cannot be the only cause," says Newton. He explains that most children with cerebral malaria do not show any physical evidence of ischemia in the brain.

Newton's group has looked at other potential causes of neurological damage and has proposed several possibilities, including raised intracranial pressure from the blocked vessels and an immunological antibody response against neurons. Several other researchers are also seeing great promise in better understanding the human immune response to malarial parasites.

The Role of the Immune System

One of those researchers, Georges E. R. Grau, M.D., professor of vascular immunology at the University of Sydney, Australia, focuses on analyzing and deciphering the biological mechanisms of the immune response. He suggests that the reason some people contract cerebral malaria while others develop only an uncomplicated form of infection pertains more to individual differences in how the immune system responds to the parasite than to the parasite itself.

"Every patient with falciparum malaria will have some parasites binding to blood vessels in his or her brain," says Grau. "But only one percent of those patients will develop cerebral malaria. Certainly there might be some parasite factors that we do not know about yet, but people with cerebral malaria have an abnormal immune response."

Grau and his colleagues have worked for many years on cultures of mice endothelial cells that have been exposed to malarial parasites. But it is not clear that the mouse variety of cerebral malaria is completely analogous to the human kind. Still, in human studies, Grau's laboratory has shown that a particular cytokine, or protein produced by immune system cells, called tumor necrosis factor (TNF) is produced in excessive amounts in humans with cerebral malaria. But Grau cautions that more than one inappropriate immune response to the invasion of malarial parasites appears likely.

"This is only one of the host responses that is abnormal," says Grau. His group has also found that those with cerebral malaria also show an elevated number of microparticles, or plasma membrane fragments released by blood cells when stimulated by TNF, in the blood. These microparticles may cause inflammation in the brain, leading to the coma, neurological damage, and death.

Newton is examining other immunologic aspects of the disease as well. In collaboration with Beth Lange, at the University of Oxford, he has found that children with cerebral malaria have an elevated number of antibodies that are associated with blocking a neuron's calcium voltage gated channels, a key part of the cell membrane that assists in the neuron's release of neurotransmitters. But how these immunological responses work to eventually lead to coma and other neurological impairment is still under investigation.

Cautious Optimism

Researchers who focus on cerebral malaria caution that we are still only at the beginning stages of the search for scientific methods to prevent the disease. Taylor states that understanding severe malarial infection is more complicated than she thought it would be. "It's a wily foe," she says. "I thought the autopsy study would have answered it all by now, and instead it's just thrown up more questions. I think the actual development of interventions will be handed off to the next generation."

But others believe some forms of more immediate relief are nearer at hand. Newton is looking into developing clinical interventions that might help prevent seizures and also ways to improve the passage of red blood cells through clogged vessels in the brain. And Grau believes that current research is providing a strong foundation for future interventions. He thinks that the scientific community is well on its way to being able to identify those who might be at risk for the more complicated forms of the disease as well as to developing a vaccine. "An anti-disease vaccine would not kill the malaria parasite completely but would, rather, prevent the abnormal set of events that follow the infection," he commented. "It's aiming at a reduction of the inappropriate host immune responses."

Wider Implications

Scientists in the field also argue that understanding the mechanisms underlying cerebral infection can play a part in understanding other diseases and neurological problems. Malaria's ability to so quickly reverse its effects is of particular interest. "Working out how a parasite can produce such a profound coma so quickly and then reverse just as quickly will be very instructive," says Taylor.

Even now, malaria is helping researchers to discover new things about the brain. By examining the brains of those who have died from the disease, a group in England led by Dr. Isabelle Medana and Dr. Gareth Turner has found that the brain has a backup reservoir of capillaries. "Normally, you can't see these capillaries because they are shut tight," says Sullivan. "But with the clogging from malaria, these reserve blood vessels open up and we now see that there are lots more capillaries in the brain itself." He argues that this finding has implications for all neurological processing and for how the brain responds to injury.

Sullivan also suggests that understanding whatever is causing the coma and overall dysfunction after malarial infection can further illuminate the ways that oxygen deprivation from blocked arteries (ischemia) causes brain damage. Newton agrees. "Cerebral malaria may serve as a useful model to study the processes set up by hypoxia, low blood flow through the vasculature."

Though malaria is not giving up its secrets easily, clinicians and researchers alike are pleased that malaria has received increased attention in the past few years. "Malaria is a preventable and treatable condition," says Newton. "It aggravates poverty and impairs the development of many regions of the world. It deserves to be high on the agenda for funding."

With the additional funding being provided for malaria research, the medical community is hopeful that it can provide answers to some of the more daunting questions about the cunning mechanisms of the disease and its human host's response. "It's a tricky disease," says Taylor. "It's going to take another five to ten years to figure out what's going on, how it does what it does." But confidence is growing that stories like Halima's will become less commonplace and that eventually there will be no need to tell such a tale again.

CHAPTER 6

Risks and Rewards of Biologics for the Brain

by E. Ray Dorsey, M.D., Philip Vitticore, M.D., and Hamilton Moses III, M.D.

E. Ray Dorsey, M.D., is a senior instructor in neurology at the University of Rochester Medical Center. His research focuses on issues at the intersection of medicine, business, and society and has been featured on National Public Radio and in the *New York Times* and the *Wall Street Journal*.
He can be reached at ray.dorsey@ctcc.rochester.edu.

Hamilton Moses III, M.D., is a neurologist, management consultant, and author. He is the founder and chairman of Alerion Advisors, LLC, and its associated Alerion Institute. Dr. Moses has advised corporations, hospitals, foundations, and governments as a partner and senior advisor with the Boston Consulting Group, was the chief physician and COO of the Johns Hopkins Hospital (1988–1994), and was interim chief of psychiatry of the Partners HealthCare System and McLean Hospital (Harvard) in Boston (1996–1998). Dr. Moses coedited Hopkins's *Principles and Practice of Medicine* and cofounded the *Johns Hopkins Medical Letter—Health After 50*. He can be reached at hm@alerion.us.

Philip Vitticore, M.D., is in his final year of neurology residency at the University of Rochester. He is a graduate of the University of Rochester, as well as the University of Rochester School of Medicine and Dentistry. He began his training in a neurophysiology fellowship in July 2007. He can be reached at Philip_Vitticore@urmc.rochester.edu.

Biologics—a new class of drugs derived from living organisms—have great potential for treating brain and other diseases. They interfere with the way a disease causes damage, rather than treating the disease's consequences, and can be individually tailored to the person taking the drug. Because biologics are created using DNA technology, rather than a uniform chemical reaction with a predictable outcome as are traditional drugs, their risks are significant and often unpredictable. The economic rewards for the biotechnology companies that develop these drugs are high, but so are the costs. The authors ask how patients, physicians, industry, and government can successfully balance these risks and rewards.

LIVING ORGANISMS, ranging from human cells to bacteria, are being used to produce new kinds of drugs that are providing promising treatments for some frustrating medical conditions. These drugs, called "biologics," rely on the insertion of a DNA sequence into living cells or organisms; the cells then grow and produce a large, often complex, protein using their own natural machinery. Biologics differ in important ways from the more traditional common forms of drugs, which are smaller, simpler molecules produced from carefully sequenced chemical reactions between inorganic (nonliving) materials.

As scientific "blue bloods," with their origins in research that earned multiple Nobel Prizes, biologics are particularly exciting to clinical researchers because they hold promise for diseases that have had few effective treatments. Already, they have become significant for several immunological, inflammatory, and neurological diseases, and the market for them has risen into the tens of billions of dollars. Despite their promise, however, the development process has also revealed serious (and unexpected) adverse effects, suggesting that caution is warranted.

Biologics for Multiple Sclerosis

The development of biologics to treat multiple sclerosis is illustrative of both the rewards and the risks of these new drugs. Treatments for disorders affecting the brain represent, so far, only a small portion of biologics developed, but biologics for multiple sclerosis have fundamentally changed clinical treatment of the disease.

Multiple sclerosis is a chronic, recurrent, inflammatory (and presumably autoimmune) disorder that results in injury of the myelin coating of nerve cells in the brain and the central nervous system. When myelin is damaged or destroyed, the ability of nerves to conduct impulses to and from the brain is disrupted, resulting in highly variable symptoms that may include cognitive changes, difficulty in walking or balance, vision problems, pain, fatigue, and bladder or bowel dysfunction. Multiple sclerosis affects both men and women, often in the prime of their lives, and is one of the leading causes of disability among young women.

Before the first biologic was developed to treat multiple sclerosis, only symptomatic treatment was available, usually anti-inflammatory medications such as steroids. The development of the biologic interferon beta-1b (Betaseron), which was approved in 1993, opened a new class of "immunomodulatory" treatment that could be taken on a regular basis to prevent relapses, slow disease progression, and potentially alter the course of the disease. Other interferons for multiple sclerosis followed, and the market for biologics for multiple sclerosis is now more than $4 billion per year.

The newest biologic for multiple sclerosis—natalizumab (Tysabri)—highlights the critical issues facing biologics as a whole. Natalizumab is a monoclonal antibody directed against a protein on the surface of white blood cells (immune lymphocytes). This antibody decreases the lymphocytes' ability to cross the blood-brain barrier and potentially cause the damaging inflammation of the myelin seen in multiple sclerosis. The biologic held enormous clinical promise in early trials and fueled the rise in the manufacturer's stock, which nearly doubled in 2004, culminating when natalizumab received initial FDA approval in November of that year.

Therapeutic biological products for multiple sclerosis

Therapeutic biologic	2006 sales ($ billions)	Approval date	Company
Interferon beta-1b (Betaseron®)	1.0*	7/23/1993	Berlex Laboratories
Interferon beta-1a (Avonex®)	1.7	5/17/1996	Biogen Idec, Inc.
Interferon beta-1a (Rebif®)	1.5	3/7/2002	Serono Laboratories, Inc.
Natalizumab (Tysabri®)	<0.1	11/23/2004	Biogen Idec and Elan Pharmaceuticals

*Betaseron® sales data are from 2005.

Therapeutic biologic products for multiple sclerosis. *Courtesy of Philip Vitticore*

At that point, the drug's performance in clinical trials appeared to be fulfilling its potential. An interim analysis of one of the major studies showed promising results, and the final results from that trial demonstrated that natalizumab decreased signs of inflammation in the brain by 80 percent (as seen on MRIs), reduced clinical relapses by almost 70 percent, and slowed disease progression by 40 percent.[1] In addition, the medication appeared to be well tolerated, and even before FDA approval, neurology clinics across the country began planning for infusion centers that, it was hoped, would administer the intravenous medication to many of the some 400,000 people in the United States who have multiple sclerosis.

The promise of natalizumab came to a screeching halt, however, with reports of three cases of progressive multifocal leukoencephalopathy (PML), a rare, often fatal, viral infection in the brain that can occur in people with lowered immune responses. This particular side effect had not been discovered in animal testing, so its appearance was entirely unexpected. While the reported cases occurred in patients who were taking both natalizumab and other immunosuppressant drugs, Biogen Idec quickly (in February 2005) voluntarily suspended the marketing of natalizumab, precipitating a dive in the company's stock price.

Now, however, natalizumab has begun a slow climb back to acceptance. On the basis of a narrower intended use and a strict patient monitoring system to track any new cases of PML, in June 2006 the FDA gave Biogen Idec approval to reintroduce the drug. Carefully chosen people with multiple sclerosis at selected medical centers are receiving the treatment, and the company's stock price has recovered a small portion of its previous losses.

The price of natalizumab, like that of many biologics, is high. Patient payments for a single monthly dose of the drug may be $1,000 or more, and health insurance does not always reimburse this. A 2006 report in the Annals of Neurology highlighted the example of a medical student at the University of California at San Francisco who was recently diagnosed with multiple sclerosis and had a poor response to beta interferon. Her physician prescribed natalizumab, but her health care plan had not yet approved the medication for coverage—leaving the student at one of the leading medical centers in the country unable to receive the medication.[2]

The Genesis of Biologics

The genesis of bioengineered drugs, such as those now available to treat multiple sclerosis, is one of the great success stories in biology and medicine, and in the interplay of those fields with commerce. The speed with which basic scientific discovery was translated into new drugs was remarkably short, less than a decade—about half the time required by most previous new classes of drugs.

Biologics—and the biotechnology companies that create them—have their origins in scientific advances made in the 1970s by researchers who learned how to manipulate genetic material. In 1972, Paul Berg, Ph.D., at Stanford University, first produced recombinant DNA, a form of hybrid DNA created in the laboratory by splicing together segments of DNA from two or more organisms or cells. Soon thereafter, Herbert Boyer, Ph.D., at the University of California–San Francisco, demonstrated the feasibility of inserting foreign DNA into E. coli bacteria. Georges Köhler, Ph.D., and César Milstein, Ph.D., subsequently discovered enzymes (called restriction endonucleases) that are able to chop up DNA at specific locations, further enabling the splicing together of recombinant DNA molecules.

Together, these discoveries made possible the development in 1982 of the first human recombinant protein, recombinant human insulin. Up to that point, people with diabetes who needed supplemental insulin had to rely on insulin obtained from cows and pigs. While that approach was effective for many people, others had allergic reactions to the foreign protein. To make recombinant human insulin, the first step is to isolate the human gene (genes are pieces of DNA) for insulin. Next, using restriction endonucleases, the gene for human insulin is inserted (spliced) into a circular piece of DNA from bacteria called a plasmid. The now recombinant plasmid (human insulin gene combined with bacteria genes) is inserted back into bacteria, where it makes many copies of itself as the bacteria divide. The growing numbers of bacteria with the recombinant plasmid produce insulin protein molecules that are gathered and purified into recombinant human insulin.

From this science, the biotechnology industry was born. Herbert Boyer helped found one of the first biotechnology firms, Genentech, in 1976. Other firms soon followed, including Amgen in 1980 and Genzyme in 1981. The recombinant DNA technology led not only to recombinant human insulin but also to recombinant factor VIII for hemophilia and recombinant erythropoietin for anemia. Since the initial production of recombinant proteins from a single human gene, applications of biotechnology have expanded to the formation of monoclonal antibodies, proteins that are directed at a specific target and underlie therapies for breast cancer (for example, trastuzumab, or Herceptin) and rheumatoid arthritis (for example, rituximab, or Rituxan). According to IMS Health (a company that provides information and analysis for the international pharmaceutical market), because of these scientific advances, the biotechnology industry is now valued at more than $50 billion and is projected to grow to $90 billion by 2009.

Rising Risks . . .

As the natalizumab example demonstrates, the industry is confronting challenges of both safety and pricing. The advantage of many of the first biologics, for example human recombinant insulin, is that they are proteins normally found in humans and therefore avoid the allergic reactions that can result from foreign proteins. But as the field has advanced, so have the technologies employed, and today many commonly prescribed biologics are "chimaeric" proteins derived from spliced DNA from more than one organism (such as humans and mice). The future will see the production of more chimaeric proteins and also the use of transgenic animals (laboratory animals with some human cells) for manufacturing new products. These novel therapies carry troublesome risks, especially that of adverse immunological reactions.

Biologics also carry idiosyncratic, sometimes fatal, risks that are impossible to predict. In 2006, six healthy volunteers in the United Kingdom nearly died after taking a particular monoclonal antibody (TGN1412), originally developed for leukemia and rheumatoid arthritis, in a Phase I trial in a small number of people to assess its safety. Despite significant investigation, the cause for the serious immune reaction has not been determined.

Because biologics are based on complex living cells or organisms, they are much more sensitive to changes in the manufacturing process than traditional drugs, which are synthesized through a controlled chemical reaction. Even a slight change in the process can cause a large problem, so biologics receive additional scrutiny from regulatory bodies. Any contaminants from the manufacturing process can produce serious adverse consequences. For example, the subcutaneous administration of a new erythropoietin product (Eprex) to treat anemia was associated with a dangerous fall in the production of oxygen-carrying red blood cells in patients. But the drop in red blood cell production decreased (or completely stopped) after changing from subcutaneous to intravenous administration of the drug.[3] According to the manufacturers, the problem occurred after the company started using a different stopper for the syringes used for subcutaneous administration.

Neurologists and psychiatrists are trying to learn from the unantici-pated adverse effects of biologics. Many of the immunologically active agents used to treat hepatitis C, cancer, and rheumatoid arthritis, for instance, produce unwanted effects in the brain, ranging from person-ality change, memory impairment, or mild depression to psychosis, seizures, and movement disorders. Fortunately, these effects are usually only temporary, but their occurrence often limits use of the biologics. Studies to understand the overlap between immunological function and the brain that occurs when these biologics are used may yield important new insights into these brain conditions.

In addition to unanticipated safety risks, producers of patented biologics face commercial risk from the possible introduction of copies. In the United States, the Hatch-Waxman Act of 1984 allows for the introduction of generic drugs for traditional small-molecule chemical products after a patent expires, if the generic can be shown to be equiva-lent to the original drug. However, the act does not cover most biologics, effectively granting many biotechnology firms monopolies without expi-ration. But competition for biologics appears to be on the horizon. The European Medicines Agency released a "biosimilar" policy in 2004. The policy would permit the introduction of similar competitor prod-ucts after the patent has expired on the initial biologic, if certain labora-tory and clinical testing studies are performed.[4] In the United States, the Access to Life-Saving Medicine Act, introduced by Congressman Henry Waxman (D-Calif.), would create a regulatory process for approval of "follow-on" (sometimes called "generic") biologics to compete with innovator biologics after patent expiration.

This issue pits the biotechnology industry—which argues that the manufacturing processes of biologics are much more complicated than those for small-molecule products and thus cannot be safely and precisely replicated—against purchasers of health care, who are looking for relief from rising health care expenses. At stake is the approximately $10 billion worth of biopharmaceuticals that are projected to lose patent coverage by 2009.[5] Resolution of this tension will tax both manufacturers and the Food and Drug Administration (FDA). Because of biologics' unique nature, companies seeking to develop copies will need to conduct more-extensive clinical testing than is typical in developing nonbiologic generics. Such trials are costly and will erode the cost/price advantage on which generics depend. Likewise, the FDA, which developed its regulatory requirements for biologics during the first twenty-five years of their availability, will be confronted by patient groups who champion greater access to medications, and to lower-cost ones.

. . . and Rewards

Biologics also offer enormous rewards for patients and companies. Because they interact with the body's own cellular processes, they interfere with the way a disease causes damage, rather than treating the disease's consequences. The effects of biologics vary from patient to patient and they can be highly targeted. As molecular genetics advances, increasing numbers of effective biologics are likely to transform treatment by supplying deficient enzymes in certain disorders, for example, Gaucher's disease, or augmenting the body's own limited supply of a protein, such as insulin in diabetes.

While the number of biologics for brain diseases and disorders currently on the market is limited, the pipeline of biologics with neurological applications is robust. Therefore, many biotechnology firms are attracting the attention of large pharmaceutical companies, which are becoming increasingly dependent on developing, or buying the rights to, and marketing new therapies as their patents on existing products expire. AstraZeneca's recent agreement to buy the biotechnology firm MedImmune for $15 billion in cash, a premium more than 50 percent of the stock price, indicates how much the biologics business is valued.

As biologics' share of the pharmaceutical portion of health care expenses grows, these therapies will continue to draw heightened public interest. Safety and manufacturing concerns, and the demand for generic competition, will likely lead to closer postmarketing surveillance. The challenge will be to continue to reward the development of pioneering

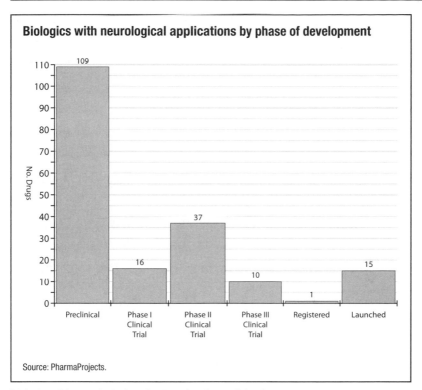

Biologics with neurological applications by phase of development. *Courtesy of Philip Vitticore*

therapies, such as new biologics to treat previously untreatable diseases, while addressing their high prices. A related challenge is to reward these pioneering innovations while exerting pressure on prices of competing products for treating the same disease. This tension is, of course, not unique to biologics, but its resolution is critical because of their especially high costs.

Issues on the Horizon

Biologics are at the nexus of many of the significant forces at work in medicine: the science, the economics, and the politics. Because no other effective treatments exist for many neurological conditions, patients and their families are outspoken. While many neurological disorders are relatively uncommon, drugs to treat them have proven to be commercially attractive, in part because of the high prices manufacturers have been able to charge for new therapies. Yet there is significant impetus to move faster and to be more productive in discovery and approval of new biologics.

Scientifically, many of the most rapidly advancing areas of neurobiology are likely to create demand for biologics. While the first applications were in immunology and hormone replacement, new target pathways particularly suitable for biologics are likely to be revealed by research in genetics (especially environmental changes to gene expression), structural biology (the chemical skeleton of nerves, synapses, axons, and receptors), and nerve regeneration and repair.

The economics of biologics will grow more challenging, especially as demand increases. The financial model of the industry has been built on aggressive pricing of drugs intended for use in a limited number of patients. Such a model will probably not continue to be viable, as pressure on price from purchasers grows and as indications for biologics extend to more patients. The consequences are likely to include greater incentives for productivity of the industry, allowing the costs of drug discovery, development, and manufacturing to be lowered.

In many ways, the politics are the most salient issue—and the most interesting. The vexing issues that surround the field display all the tensions at work between industry and its regulators, on the one hand, and physicians and patients, on the other. A current example is pressure for early access to new biologics, before FDA approval.

Conventionally, access to any drug that has not yet been approved is restricted to patients who are enrolled in controlled clinical trials. However, many now advocate modifying this restricted access to new treatments that are in development. One particular remedy, urged by patient advocacy groups, is the patient-sponsored trial, one that is paid for by the patients themselves. This idea was borrowed from the field of cancer treatment, where it was first used for the treatment of rare tumors, especially in children. Because of the infrequency of such tumors, the FDA and pharmaceutical companies were generally supportive of this practice and afforded considerable flexibility in obtaining "compassionate use" drugs. Also, insurers were often willing to offset the cost of such therapy, just as they had done historically for experimental cancer chemotherapy. Neurological patients would like to see the same kind of flexibility available to them. Several lawsuits currently in the courts argue for a "right to access"; the rulings in this litigation will no doubt have particular bearing on the future of biologics.

From a therapeutic standpoint, the future of biologics appears robust. Researchers are likely to develop new biologics that address many conditions, often in ways that are tailored to the specific pathology of a disease and to the person taking the drug. But, despite its noble scientific roots, the future of biologics will increasingly be determined by debates in the economic and political spheres. Like health care as a whole, the field of biologics will have to justify its high prices and demonstrate the economic value of its products. Because of its cutting-edge science, biologics will be at the leading edge of political discussions about the science that should be funded, the types of therapies that will be developed, and the people who will initially have access to them. Developing innovative resolutions of these debates will require the collective input of patients, physicians, industry, and government.

CHAPTER 7

"Cosmetic Neurology" and the Problem of Pain

by Anjan Chatterjee, M.D.

Anjan Chatterjee, M.D., is an associate professor at the Center for Cognitive Neuroscience, University of Pennsylvania School of Medicine. His research focuses on human cognition, especially language and meaning, visual aesthetics, and visual, spatial, and temporal processing. He is also interested in the ethical dilemmas arising from advances in clinical and basic neuroscience. He can be reached at anjan@mail.med.upenn.edu.

Few people would argue against treating the traumatic psychological effects of war or violence. But what about taking a drug to lessen the pain of our common daily struggles, such as the end of a relationship or anxiety about one's job? Is this a "cosmetic" enhancement of human life, even a danger to character, or is it an ethical choice? For guidance, the author looks to the history of treating physical pain and argues that, despite growing knowledge of the biological basis for psychological pain, many find it hard to find a consistent principled position when it comes down to specific instances of alleviating human suffering.

We are all familiar with—and many are troubled by—athletes who use medications, legal or otherwise, to enhance their performance. This practice is an early indication of a much larger trend. As neuroscience advances, we are getting better at treating cognitive and emotional disorders, and we are also learning how to improve cognition and modify emotions in basically normal, healthy people—for example, by increasing alertness or lessening fear. I have coined the term "cosmetic neurology" for this practice.[1]

Cosmetic neurology raises four major ethical concerns. First is a concern about safety. We weigh the potential risks and side effects of a new medication for a disease against the potential benefits. In health, are any risks worth taking? For example, musicians often use beta-blockers to dampen tremors and anxiety associated with public performance. Occasionally, however, beta-blockers are associated with a life-threatening anaphylactic (allergic) reaction in which a person's blood pressure drops and breathing stops. Is the better concert worth this risk? Second is a concern about distributive justice: If cosmetic neurology succeeds in making people smarter and happier, will these enhancements be available disproportionately to the affluent? Third is a concern about coercion. Will healthy people be or feel forced to take such medications, either because it would serve a greater good (for example, airline pilots being required to take a drug to increase alertness if that made flying safer) or because of competitive pressures?

Finally—and this is the focus of this article—ethicists and others have expressed a subtle but deep concern about ways in which manipulating our emotional lives might erode character, both individually and communally. This was a fundamental concern raised by the President's Council on Bioethics and highlighted in its 2003 report, *Beyond Therapy*.[2] If, as many religions and philosophies argue, struggle and even pain are important to the development of character, does the use of pharmacological interventions to ameliorate our struggles undermine this essential process?

The widespread practice of cosmetic neurology seems inevitable, and resolving this concern will not be easy. Many people share an underlying discomfort with how things might play out. But when we consider specific instances of using a medication to affect emotion and treat psychological pain, it's not always clear that an individual would be better off without the drug. What some might see as a dubious or even dangerous enhancement, others believe is an ethical means of relieving suffering.

In my view, the history of the treatment of physical pain, including "natural" pain, anticipates the treatment of psychological pain. Similar tensions are certainly at play. This claim is predictive, not prescriptive—I am neither advocating nor decrying the use of cosmetic neurology. I am, instead, pointing out how deeply difficult it is for anyone, ethicists included, to adopt a consistently principled position on the problem of pain.

The Varieties of Cosmetic Neurology

The pharmaceutical and other interventions that we place under the heading of cosmetic neurology target the human motor, cognitive, and emotional systems. The functioning of our motor system can be enhanced by influencing the cardiovascular, peripheral motor, and central nervous systems. For example, both the hormone erythropoietin and the drug sildenefil can be used by athletes to increase the oxygen-carrying capacities of their blood and provide better endurance, and drugs that act on receptors for the neurotransmitter dopamine may very well improve our ability to learn new motor skills.

Attention, memory, and learning can also be altered in healthy people, sometimes with drugs developed to treat a disease and sometimes with treatments created specifically to enhance cognitive abilities. Medication that boosts the effects of the neurotransmitter acetylcholine, used widely to treat symptoms of Alzheimer's disease, has been shown to improve alertness and attention in healthy people, as has modafinil, a drug used to treat sleep disorders such as narcolepsy. Stimulants such as atomoxetine, which is used to treat attention deficit disorder, are also likely to improve levels of alertness in normal individuals. In addition to these currently available drugs, new classes of drugs that could be used as cognitive enhancers are being investigated. Some of these promote structural changes in neurons that accompany the acquisition of long-term memories. These drugs—molecules with names like "ampakines" and "cyclic AMP response element binding protein modulators" that may one day sound as familiar as "statins" does today—are designed not to treat pathology but to exploit normal biological processes, with the hopes of improving memory.

Finally, and most relevant to this discussion, our understanding of the brain basis of emotions continues to grow, as does our ability to modify the systems related to various emotions. Take, for example, fear. It is clear that a brain structure called the amygdala, located within the temporal lobe, is involved in regulating the effects of fear and our responses to it.[3] The amygdala receives signals from pain pathways, from higher-order perceptual processing areas, and from the hippocampus, an area long known to be essential to the formation of memory. The amygdala, in turn, sends signals to the hypothalamus, which regulates stress hormonal responses, and to areas of the brain that regulate arousal, such as the locus ceruleus. Thus, the amygdala is a critical structure that controls our experience of fear and colors our perception and memory of fearful events. As shown in recent experiments,[4] the effects of fear on memories can be dampened by local infusions of beta-blockers (long used to treat high blood pressure), thereby helping with symptoms of anxiety.

The drugs called selective serotonin reuptake inhibitors (SSRIs) are commonly used to treat depression and anxiety, but they could have wider applications. For instance, research on primates has shown that infant monkeys that have been abused have lower serotonin levels in their brains than those who have not been maltreated, and those infants with the lowest levels are more likely to become abusive adults. Humans with a specific form of a serotonin transporter gene have abnormal amygdala activity and are especially prone to fear and anxiety, as well as to the effects of abuse. In healthy people, SSRIs promote "affiliative behavior," or friendly positive behavior toward others. So one might argue that these drugs should be used even more widely than is the current practice.

Just around the corner are several new ways of potentially controlling affective (mood) states by regulating neuropeptides, small proteins in the brain that influence how information is processed and that can be linked to quite specific behaviors. One such neuropeptide, corticotropin-releasing factor, seems to influence neural changes produced by ongoing stress. These changes include lowered levels of neurotransmitters that influence attention and mood, such as serotonin, epinephrine, and dopamine. Blocking corticotropin-releasing factor, which would lower glucocorticoid levels, might blunt these long-term effects of stress. The hormone oxytocin might be used to induce trust. Other neuropeptides, such as substance P, vasopressin, galanin, and neuropeptide Y, are also being studied as potential targets for treating the brain's emotional functions.

Psychological Pain

Most reasonable people agree on the urgent need to ease the psychological burdens imposed by significantly traumatic events. For instance, thousands of young men and women experience varying degrees of post-traumatic stress disorder following military service, and many of them fall through the cracks in our society. Few people would argue against treating such individuals, in the same way that few would argue against treating their physical ailments—even if, in practice, we fall short of treating either type of affliction.

But what about less-severe traumas, or even the challenges of everyday life? Preliminary research suggests that beta-blockers may prevent post-traumatic stress symptoms when given to people who have gone to the emergency room after a car accident. In addition to dampening the emotional effects of memories after they form (retroactively), such medications could presumably be used proactively, when the memories are first encoded. If they are proven effective, and more such treatments become available soon, how widely would they be used? We could expect people to employ them for all sorts of "normal" traumas, such as remorse over wrongdoing or breakups in relationships and the other losses and disappointments that seem integral to our existence as humans.

But what would be the long-term consequences of flattening these bumps in the road? Do we need the experience of pain to develop character? Beyond individual development, what is the role of pain in binding us communally? Researchers have learned that empathy for the experience of pain in others may be made possible by the observer's own neural pain circuits, particularly through the anterior cingulate and the insula. If pain circuits are chronically dampened, would a person still be capable of empathy? Would our society be less caring of marginalized groups, such as those with mental illnesses and other disabilities?

Anesthesia for Physical Pain

Approaches to the problem of pain have historic precedents in the treatment of physical pain, particularly the use of anesthesia to ease the pain of surgery and of childbirth.[5] In October 1846, William T. G. Morton, a dentist in Boston, gave the first public demonstration of the use of anesthesia in surgery; and on January 19, 1847, James Young Simpson, a Scottish obstetrician, used ether to facilitate the delivery of a child by a woman with a deformed pelvis. Simpson became a forceful advocate of the wide use of anesthesia for childbirth, a practice that was extremely controversial at the time. Medical discussions about the benefits and risks of anesthesia played a relatively minor role. At the heart of the objections was the social construction of the meaning of pain. In this light, treatment of pain was objectionable on three grounds.

Pain as Natural

First, don't mess with Mother Nature. Some pains are natural. We should not be meddling with the natural course of things, since we are not wise enough to predict the unintended consequences of our meddlesome ways.

From the very beginning, some obstetricians objected to the use of anesthesia for childbirth on grounds that childbirth was natural and interventions such as the use of ether or chloroform invited medical disaster. As physicians took the possibility of side effects seriously and tried to mitigate them, safety became less of a concern and the popularity of anesthesia continued to rise. But the appeal of all things natural resurfaced with force in the mid-twentieth century, when Grantly Dick-Read promoted the natural childbirth movement and Fernand Lamaze published *Painless Childbirth*. A professional rivalry between Read and Lamaze increased the public's awareness of the possibilities of natural childbirth, and in 1956 Pope Pius XXII gave a special address on the moral value of natural childbirth, giving these approaches spiritual weight. This address coincided with a period in which the public was losing confidence in medicine's ability to alleviate illness and pain.

Pain as Punishment

Second, spare the rod, spoil the child. Sometimes we deserve to be punished. Pain, as a form of punishment, structures individuals and orders our society. To mitigate pain would make for a society of sinners as we succumb to our lesser angels.

Pain plays a central role in many religious traditions and is often viewed as an acknowledgment of human imperfection. The link between the pain of childbirth and punishment is made explicitly in Genesis 3:16: "Unto the woman he said, I will greatly multiply thy sorrow and thy conception; in sorrow thou shalt bring forth children." The notion of pain and suffering as deserved is evident in other traditions as well, and self-infliction of pain as an act of atonement remains prominent in Christian, Muslim, and Hindu belief. Similar views of the role of pain as punishment can be observed in secular institutions. For many years, brutal public executions were sanctioned to serve as both public entertainment and education. Humanists debated the use of pain in legal systems, and despite prison reform movements, the general impression that social order would disintegrate if the law did not use its authority to punish and inflict pain remains robust. In this view, relieving deserved pain would be hubris at best, and more likely an invitation to chaos.

Pain as Progress

Third, no pain, no gain. Learning to cope with pain strengthens and deepens us. Mitigating pain could cheapen us, individually and communally.

In spiritual frameworks, pain serves as a vehicle for transcendence. The Christian symbol of the cross exemplifies the link of sacrifice and salvation. In secular views, pain builds character. Writers have explored the experience of pain and suffering as integral to larger-than-life characters in literature, such as Hamlet and Faustus. F. Scott Fitzgerald claimed, "You especially have to hurt like hell before you can write seriously,"[6] a sentiment echoed by others who have linked pain to creativity.

Pain also serves to strengthen social bonds. Religious views that a God that punished also healed meant that communities rejoiced together in that healing. The pain of childbirth and the real possibility of death meant that neighbors and family and friends supported the event in a way that often formed lifelong social bonds. As childbirth moved from the home to hospitals in technically developed countries, many of the rich social and cultural traditions were reduced to ritualistic baby showers. Perhaps in partial reaction to this sterile approach, in 1948 the Royal College of Obstetricians and Gynaecologists found that half of 15,000 British women interviewed preferred delivery in their home, which meant with a midwife and without anesthesia. Such attitudes fluctuate over time, but concerns about the "medicalization" of childbirth remain germane today. In 2002, the *British Medical Journal* devoted a special issue (vol. 324, April 13) to discussions of medicalization trends in general, and noted with some alarm the rise in Cesarean section deliveries and the inappropriate use of fetal monitoring.

Reinterpreting Physical Pain

Despite the various ways in which the treatment of physical pain was (and sometimes continues to be) viewed with mistrust, the use of medications and anesthesia for pain management is now widespread. Indeed, a growing international consensus calls for effective treatment of pain as a fundamental human right.

Two reinterpretations of pain facilitated this change. First, the classification of pain as a biological phenomenon diluted the impact of religious interpretations. When Simpson began administering ether to women giving birth, he emphasized that the pain of childbirth was a consequence of anatomy and not divine wrath, even suggesting that God was an advocate of anesthesia, as evidenced by his putting Adam into a deep sleep to extract the rib that would become Eve. In the early twentieth century, the physiologist Sir Charles Sherrington observed that complex behavior could be analyzed as a set of coordinated neural reflexes. The discovery of various sensory receptors and the signaling of pain by specific neural pathways made explicit the possibility that pain could be altered, lessened, or even blocked. This mapping of physical pain onto its biology helped frame the treatment of pain as simply one more mechanical manipulation.

Second, in the late nineteenth century, the public attitude toward pain and suffering of all kinds shifted. William James wrote, "A strange moral transformation has, within the past century, swept over our Western world. We no longer think that we are called on to face physical pain with equanimity. ... The way our ancestors looked upon pain as an eternal ingredient of the world order, and both caused and suffered it as a matter-of-course portion of their day's work, fills us with amazement."[7] This transformation occurred in the setting of broad-based humanitarian movements dedicated to the relief of suffering in many forms. Women's suffrage and abolitionist, prison reform, child labor reform, and antivivisectionist movements gathered force during this period. Alleviating physical pain fell naturally within the purview of these humanitarian efforts.

Psychological Pain as Biological

In exploring ethical concerns about anesthetizing psychological pain, the differences between it and physical pain are less relevant than the similarities of their underlying neurobiology and, therefore, of their socially constructed meaning.

Physical pain produces neural responses that fractionate into three components.[8] First is the sensory experience itself, which is processed in the brain by parts of the somatosensory cortex and by a deep structure called the thalamus. Second is the subjective sense of "unpleasantness," which is often, but not always, correlated with the intensity of the pain sensation. This subjective sense of unpleasantness is accompanied by neural activity in the insula, which controls our autonomic nervous system, and the anterior cingulate, which integrates the cognitive experience of pain with its emotional aspects and establishes priorities for responding to the pain.

Finally, pain also produces what is called the "secondary pain affect," which refers to emotional feelings about the long-term implications of having pain that can last long after the inciting physical pain. The neural basis of the secondary pain affect is not well understood, but it is thought to emerge from an interaction of the anterior cingulate, insula, amygdala, and prefrontal cortex. These same parts of the brain are also part of the distributed network that coordinates emotional distress and its interactions with our cognitive systems. We can see how physical and emotional pain converge in the brain.

As such understanding of the neurobiology of emotional systems deepens, treatment of psychological pain becomes easier to view as a mechanical rather than a metaphysical manipulation. Similarly, it is hard not to see such intervention in humanitarian terms, rather than as a cosmetic "enhancement." In addition to the wave of post-traumatic stress disorder that will soon be upon us as a result of veterans returning from the Iraq war, some estimate that more than a quarter of the American population suffer from affective or addictive disorders.[9] How could anybody object seriously to the alleviation of this suffering, even if it means that others might take the same pills for more trivial reasons?

The Ethical Dilemma of Pain

We face a fundamental problem in trying to establish a coherent ethical position on ameliorating psychological pain. The general unease shared by most people about ubiquitous treatments of such pain coexists with competing and conflicting attitudes about specific situations.

We can worry about loss of character individually and communally at the same time that we are willing to frame psychological pain in biological terms or consider its treatment a broad humanitarian goal. We might share the general sense that some things are best left alone, but we are unlikely to agree on which specific things are best left alone. We might share the general sense that pain serves a purpose in establishing order, but we are unlikely to agree on which pains can be justified and for whose version of order. We might share the general sense that pain can be a vehicle for character development, but we are unlikely to agree on which specific pains are worth enduring for a greater good. If the same person weighs these considerations differently for each situation, and changes the relative weightings of these considerations for the same situation at different times, then how could one possibly have a coherent reflective position?

The holding of contradictory beliefs with unstable relative weightings makes it extremely unlikely that ethicists, as a group, will be able to speak with one voice. As a result, they will be unlikely to shape social norms that could guide a coherent practice of anesthetizing psychological pain or provide a basis for public policy. Without such restraint, aspects of cosmetic neurology, at least the practice of modifying emotional systems in basically healthy people to lessen suffering, great or small, will flourish.

My claim is not that everyone will use medications to alleviate the bumps and bruises of everyday living. It is that more and more medications will become available for this purpose and at least some people will find it ethically acceptable to use them. Anesthesia for childbirth is available to virtually everybody in technically developed countries, but some choose to use it and some do not. The extent of general use of interventions for psychological pain will fluctuate with people's faith or frustration with science, technology, and medicine.

CHAPTER 8

When Music Stops Making Sense

Lessons from an Injured Brain

by Petr Janata, Ph.D.

Petr Janata, Ph.D., is an assistant professor of psychology at the Center for Mind and Brain, University of California, Davis. His research focuses on how the basic neural systems that underlie perception, attention, memory, action, and emotion interact in the context of natural behaviors, with a particular emphasis on music. He can be reached at pjanata@ucdavis.edu.

In 2004, Ian McDonald, M.D., a British neurologist and amateur classical pianist, experienced a stroke that damaged a fairly small area of his brain. As a result, he temporarily lost his ability to read and play music from a score, as well as to appreciate music emotionally. What does his experience, and that of other people whose musical abilities have been affected by brain damage, teach us about how the brain binds together what we perceive into a seamless flow?

AS WE INTERACT WITH OUR ENVIRONMENT in the course of our daily lives, most of us take for granted the smoothness with which events blend together in a single flow. Our brains stitch together what might otherwise seem to be a disjointed sequence of individual events into a coherent stream of information about the world around us. Language is an obvious example, of course, as letters are bound together into words, words are grouped into phrases and sentences, and sentences form paragraphs and discourse. But music, too, relies on our ability to bind together smaller elements into larger structures. Individual notes are strung together to form melodic motifs. Melodic motifs are used to shape phrases, which in turn form songs or even movements of symphonies. Notes sung simultaneously by different voices or played by different instruments are combined to create harmonies. How our brains bind together these discrete pieces of auditory information to create the experiences of hearing, remembering, or performing music is at the heart of a recent surge of neuroscience research aimed at understanding this ubiquitous human phenomenon.

Despite music's central role in human cultures around the world, and its potential to help unlock the mechanistic secrets of the brain, its arrival on the scientific scene is rather recent. Nonetheless, exploring music's basis in the brain can help shed light on a remarkable human activity that has been a part of our social and cultural fabric for millennia. Moreover, while the relegation of music to scientific second fiddle is understandable, we should not minimize the role that music can play in our broader understanding of how the brain works.

Translating a Musical Score into Action

Understanding how the brain accomplishes music is likely to enhance our understanding of the brain's inner workings for the simple reason that musical behaviors include the same elements of perception, action, emotion, and other mental operations as so many other kinds of behavior. To get a handle on how music might be processed by the brain, scientists use two primary methods: case studies of people who have suffered some form of brain damage and neuroimaging studies to measure physiological changes in the brains of healthy people while they perform various tasks related to music, such as remembering short melodies or reproducing rhythms. Studying people with an injury to the brain provides a particularly fascinating window into the workings of the brain because it helps researchers understand general principles of brain organization while taking into account variations in individual cases.

Following damage to the brain, such as the death of tissue after a stroke, a person is likely to experience deficits—whether in perception, movement, attention, or memory—that make performing previously routine behaviors and tasks more difficult. If neurologists carefully identify these deficits, they can pinpoint the specific mental operations that are impaired. When multiple patients who have damage to the same brain area experience the same deficits, the neurologist can ascribe the underlying mental operations to specific brain areas. For example, damage to the left side of the frontal lobe of the brain in a region known as Broca's area results in Broca's aphasia, the inability to produce sequences of speech sounds and words.

Closer scrutiny of individual patients presents a more intriguing picture, however. Often some highly specific functions are lost but others are spared, despite the brain damage being widespread or occurring in a region commonly associated with a multitude of functions. In this article, my launching point for exploring music in the brain is the

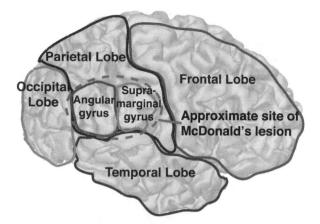

Figure 1: The right side of the brain, showing the location of the parietal cortex and the damage from Ian McDonald's stroke. *Courtesy of Petr Janata*

remarkable personal account of one such patient, Ian McDonald, M.D.[1]

McDonald was a British neurologist with an avocation as a classical pianist. He was a skilled musician who spent much time playing the piano, both alone and in small chamber music groups. But as a result of a stroke in 2004, he lost his ability to read and play music from a score, as well as to appreciate it on an emotional level. Fortunately, he documented both his symptoms and his recovery. His report, taken together with other similar case studies, provides insight into the role that the brain's parietal lobes play in transforming information from one form into another and in binding together the stream of events into a continuum of meaningful experience.

Initially, McDonald did not realize he had experienced a stroke, but after becoming aware of various difficulties with previously routine tasks, he sought a medical evaluation. The resulting exam and tests showed that the stroke affected a relatively circumscribed area of his brain, two folds in the surface of his right hemisphere that are called the right supra-marginal and angular gyri. These folds are part of the parietal cortex, as shown in the figure above.

On the mechanistic level, McDonald reported considerable difficulty in reading a musical score. He suffered from "musical alexia"—an inability to read music—that was so pronounced he had trouble reading

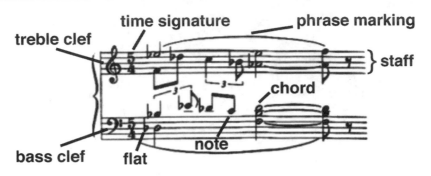

Figure 2: The components of a musical score. *Score © Gardner Read*

even simple melodies. He made clef errors, meaning that he interpreted music written in the bass (bottom) clef—generally played by the left hand—as being written in the treble (top) clef, generally played by the right hand. He also had a hard time reading the notes written above or below the staff, the set of five lines on which notes are marked. Fluent reading of a musical score requires more than reading individual notes on a page; it also requires translating the notes into actions. So it is interesting that McDonald also had problems moving his hands to the appropriate places on the keyboard, especially when larger jumps across many piano keys were involved.

Remarkably, even though he was unable to read the score fluently, he was able to give letter names to the notes in the score (A, B, C, and so on) and then use that information to pick out single notes of a melody on the keyboard, one at a time. Thus, it wasn't that the score made absolutely no sense to him; rather it was that he couldn't read it in the way to which he was accustomed, by translating the notes directly into action without having to think of their names. What might a neuroscientist see in McDonald's ability to translate the information in a musical score into appropriate key presses on the piano using the more circuitous and tedious process of verbal labeling?

To understand the implications, we must first understand what cognitive neuroscientists mean by the term "representation." This word simply reflects the idea that things in the external world exist separately in the external world and in our minds. In the case of reading music, physical information about a written musical note (a black oval with a vertical line extending from the side of it) can be present in the brain at multiple levels of abstraction. These mental representations are embodied in patterns of activity among groups of neurons in different parts of the brain. For example, neurons at early stages of processing in the visual cortex will respond to the physical properties of the note—the size of the black oval and the line—thus representing those physical aspects. But a higher stage of processing is required to interpret (and therefore represent) the particular combination of physical features that make up the note.

So McDonald's ability after his stroke to translate notes on a page via a different method suggests that information about a musical score can be represented in different ways in multiple brain regions. It also indicates that these representations can be combined flexibly to meet different goals, for example, naming the notes on a page versus playing them with a desired phrasing.

If a musician wants to spell out the notes that form a chord or a melody, for example "C-E-G," the visual image of each note on the page must be bound together with the verbal label that corresponds to that note. Associating a note on a specific line, or space between two lines, with a letter name isn't enough, because the same line in the treble (top) or bass (bottom) clef corresponds to a different note. The problem is further compounded by the key signature—the sharps and flats—which typically appears only at the beginning of the staff, next to the clef sign. All of these pieces of information together provide the context that shapes how the brain translates a circle appearing on a line—a note— into a letter label. This context must be held in the musician's working memory, a short-term buffer that the brain uses to maintain and compare pieces of information.

Verbal labeling takes time, but musicians usually translate the visual image into movements more quickly. The same pieces of visual and contextual information that go into verbal labeling become bound instead to action plans that move the hand and fingers to appropriate locations on the piano or other instrument. Because it was this ability to translate a musical score into action that was damaged after McDonald's stroke, and because the stroke affected the angular and supramarginal gyri in his right parietal hemisphere, we can logically suspect that these parts of the brain are involved in binding together the mental representations of the notes and the actions.

Researchers have learned recently that corresponding regions in the left hemisphere are also involved in reading music. But damage to that hemisphere has the opposite effect than the one that resulted from McDonald's stroke. Daniele Schön, Ph.D., and his colleagues at the universities of Trieste and Padova studied a professional musician who had a stroke that damaged her left temporoparietal area.[2] In direct contrast to McDonald—who could name notes but not read and play them—she was unable to name the notes of a melody written in the bass clef, even though she could read the melody from the score flawlessly when playing it on the piano. Similarly, she could play notes whose names were spoken to her, but she had pronounced difficulty in writing those same notes. So in this case, the ability to bind note names with their visual symbols was impaired.

A similar failure of the note-naming system was evident in another patient described by Lisa Cipolotti, Ph.D., and her colleagues at University College, London.[3] This patient had left hemisphere damage and considerable language impairments (aphasia), but virtually no musical impairments with the exception of the inability to name written musical notes or produce a note from a verbal label. In this particular case, the patient had trouble reading letter names and comprehending letter sounds, regardless of whether they were part of words or notes in a score, illustrating how a common mental representation of letters is likely involved in both language and music.

Neuroimaging the Healthy Brain

These observations in people who have experienced a brain injury are complemented by observations that were made in neuroimaging studies of people without any brain damage while they read music.[4, 5] These studies, taken together with the location of McDonald's brain lesion from the stroke and his impaired score-reading, strongly implicate the right parietal cortex as a nexus in the multifaceted process of reading a musical score.

In order to isolate those regions involved in a particular mental process, neuroimaging studies carefully compare two experimental tasks that differ in only one critical aspect. In one key study, Justine Sergent, Ph.D., and her colleagues at the Montreal Neurological Institute used a series of tasks to identify brain regions involved in the mental processes associated with reading and playing melodies. To identify what parts of the brain are involved in playing simple musical scales, Sergent and her team subtracted the activity in the brain of a person listening to musical scales being played by someone else from the activity when the same person both plays a scale and hears it played. The main difference between these two conditions was the act of playing, because in both cases the subjects heard the scale.

Next, to get at the more complex processes associated with reading a musical score, they compared brain activity caused by reading a score with that caused by viewing and responding to a simple pattern of dots on a screen. The idea was to figure out where in the brain a musical score starts to assume meaning, once the lower-level perceptual processing of the notes as simple dots has been accomplished. They also compared the brain activity elicited by both reading and hearing a musical score with that involved in simply reading the musical score, so they could identify those brain areas responsible for hearing the music.

As a result of these types of comparisons, Sergent and her colleagues learned that the visual processes involved in reading a musical score engage parietal regions at the border of the parietal and occipital cortices, while the auditory processes engage the supramarginal gyrus. Following up with even more specifically targeted comparisons helped them to identify areas that bind together representations of the sensory information—for example joining the position of a note on a staff, or the letter name for that note, with the appropriate actions, such as the hand or finger movements necessary for playing the note. As a result, they determined that parts of the superior parietal lobule, along with a broader cortical network of motor areas, are involved in these interactions.

More recently, Daniele Schön and his team were interested in finding what brain areas specifically represented the mapping between reading a musical score and the musical action. They sought to isolate the specific areas involved in directly converting a score into movements, using an experiment that compared playing in response to letter names versus playing in response to Arabic numerals that were used to identify the different piano keys. Their study identified specific sites in the right superior parietal lobule and the inferior parietal sulcus (a region that forms the upper border along the angular and supramarginal gyri) that were more active specifically during score reading.

Emotion and Time

As mentioned earlier, McDonald's piano playing was profoundly affected not just in terms of mechanically reading and playing the piano but on an emotional level as well. He observed, "I seem to have lost the ability to follow from one chord to the next. The music did not seem to 'make sense,' that is, there did not seem to be the necessary or implied connection between one chord and the next of which anyone familiar with this music is aware." Commenting further on this lack of cohesion between successive chords or notes, which made it difficult to discern the emotional intent of the composition, he wrote, "The succession of notes, even when correct as a result of having designated them by letters, made no musical sense: the emotional content the sequence of notes should convey, which determines phrasing and expression, was completely absent."

These statements by McDonald provide two intriguing insights into the functions of the brain regions that were damaged. First, they suggest that these areas not only bind together features of an object or event at a single point in time—for example, binding the image of a note with its letter name—but also bind information over a period of time, such as connecting a series of notes into a coherent musical phrase. This makes sense because numerous studies have found that these areas of the brain are activated during tasks that require maintaining a sequence of items in working memory. Keeping contextual information in mind, such as which clef the notes on the staff belong to or which key they are in, also depends on working memory.

More intriguing is the idea that this area of the parietal cortex is somehow involved in an emotional response that arises from the cohesiveness of a series of musical notes and phrases. This suggestion is surprising because emotion is typically associated with other parts of the brain, such as the limbic system or parts of the prefrontal cortex. Thus, one has to find a way to reconcile the more mechanistic functions of the parietal cortex—for example, sensorimotor binding and working memory—with the more abstract and ephemeral experiences of emotion and meaning.

Ultimately, all behaviors and their emotional manifestations depend on coordinated activity in multiple brain areas, so a possible solution to this conundrum presents itself if we consider the role of the parietal cortex in the context of the broader brain networks of which it is a part. McDonald's observations suggest that part of the parietal cortex might be a gateway through which the emotional systems of the brain find out whether a sequence of events and actions merges into a coherent experience. The work of Marcus Raichle, M.D., and his colleagues at Washington University in St. Louis shows that the supramarginal and angular gyri belong to reciprocally active networks.[6] The supramarginal gyrus appears to be part of a network that supports directing attention outward to objects and events in our environment, maintaining information in working memory, making decisions, and implementing actions. In contrast, the angular gyrus appears to be part of a network that is more active when a person's thoughts are directed inward, as when evaluating how one feels about something, or when forming larger-scale action plans. It might, therefore, be critical in giving music its emotional meaning.

The Multiple Roles of the Parietal Cortex

Cognitive deficits arising from brain damage caused by stroke rarely affect only a single mental process or ability such as language, semantic knowledge, memory, or attention. The musical deficits in the neurological cases discussed earlier were generally part of a broader spectrum of compromised abilities. Although his language skills were largely unaffected, McDonald experienced not only musical impairments but also a variety of other difficulties that required the combination of spatial, visual, and temporal information or the comparison of information in working memory. For example, he became unable to navigate in a foreign city using a map, and he had difficulty crossing the road because of problems with the spatiotemporal abilities necessary to judge the direction and speed at which cars were traveling. He also suffered from dyscalculia—an inability to do mathematical calculations in his head—when he tried to convert the value of one currency into another. Like binding together notes into a musical phrase, such skills also require maintaining and combining multiple pieces of information across time and are known to involve the parietal cortex.

Some of the other people with musical alexia who have been studied by researchers also experienced other deficits, typically in their language abilities, as well as in other musical skills in addition to reading a score. In one case, the patient's ability to reproduce a rhythm or tap along with one was impaired, but her ability to recognize familiar melodies and discriminate between melodies was preserved. In contrast to McDonald's right hemisphere lesion, the lesion in her case was on the left side of the brain, spanning the area from the auditory cortex to the angular gyrus, perhaps explaining why she had greater language difficulties.

Although in this article I have focused on only a few cases of musically talented people with brain damage, the patterns of deficits that result from damage to the same general brain region are striking. On the one hand, certain common principles emerge. For instance, information is bound across time and perceptions are bound to actions. This suggests that the parietal cortex plays a critical role in binding together different kinds of mental representations. However, the patterns of highly specific deficits are not consistent, which raises an interesting question about the modularity of functions in the brain and whether separate modules for music and language exist. Cases in which language is impaired (aphasia) but musical ability is not, and vice versa, support a modular view. But perhaps such dissociations between the two abilities are, instead, simply an idiosyncratic consequence of a person's individual life experiences. In other words, to what extent is the organization of the parietal cortex a consequence of a lifetime of experience and particular skills, each of which requires putting different pieces of information together in different ways? To the extent that various people become expert in different behaviors, or implement the same behaviors and skills in different ways, the details of their brains' parietal topography may also vary.

Let's imagine for a moment that, rather than playing the piano by reading a musical score, McDonald routinely improvised and played by ear. Or imagine that he made music proficiently in all these ways. Would the stroke have affected these music making abilities equally, or would he, nonetheless, have experienced only musical alexia? My guess is that the music still would have stopped making sense to him and would have lost its emotional richness, even if he was playing by ear. Although this type of deficit seems qualitatively different from the inability to put together specific pieces of information, such as the positions of notes on a staff with the appropriate finger movements, it does suggest that cohesion among events is an important component of our emotional responses to those events.

One way to try to dissociate the emotional components from the more mechanistic aspects of binding information might be research using transcranial magnetic stimulation. This technique employs pulsing strong magnetic fields above specific brain areas in order to create temporary lesions. I would predict that stimulating the angular gyrus would result in a transient loss of the sense of emotional meaning without affecting musical score reading, whereas stimulation of the adjoining supramarginal gyrus might have the opposite effect.

McDonald's experience and the other case studies I've described highlight the significance of the parietal cortex as a sort of switchboard and sequencer, tasked with selecting and binding together those pieces of information that are required for fluid sequencing of actions. Twenty years ago, parts of the parietal cortex were recognized primarily for their critical role in visual attention, and neurological deficits in that area were associated with visuospatial neglect—the tendency to ignore the side of the space outside oneself that is opposite to the side of the lesion. With the advent of neuroimaging, it has become evident that the parietal cortex is involved in the more general directing of attention and that the process of selecting objects or locations on which to focus attention is closely related to working memory, which holds multiple representations in mind in the service of achieving some sort of goal.

For those who would like to ascribe a single function to a single brain region, this view of the parietal cortex as a multipurpose dynamic router might be too nonspecific and unsatisfying. But others, I among them, view the core mental operations at work as selecting, binding, and sequencing information. For us, the joy comes about in trying to understand the neuroanatomical and functional details of how and why specific pieces of information—such as pictures of musical notes, names of notes, and actions associated with notes—become bound in the way that they do.

Studying the functional organization of the brain provides constant reminders that the organization and manipulation of perceptions, thoughts, and actions are rarely as straightforward as we might hope or imagine. As we examine individual cases of people with brain damage, such as McDonald, and review functional neuroimaging studies that purport to examine unrelated unique phenomena, we are reminded that the brain's processes underlying highly specific behaviors and skills are actually more intertwined than was at one time believed. By taking into account similarities across the multiple behaviors at which humans excel, such as music, language, mathematical reasoning, and planning, we stand a better chance of elucidating the universal principles of human brain function.

CHAPTER 9

Seeking Free Will in Our Brains

A Debate

by Mark Hallett, M.D., and Paul R. McHugh, M.D.

Mark Hallett, M.D., is chief of the Human Motor Control Section at the National Institute of Neurological Disorders and Stroke (NINDS), National Institutes of Health. His research focuses on the physiology of human movement and movement disorders. He is currently editor in chief of *Clinical Neurophysiology* and an associate editor of *Brain*. He was the clinical director of NINDS until July 2000 and is past president of the American Association of Neuromuscular and Electrodiagnostic Medicine and the Movement Disorder Society and past vice president of the American Academy of Neurology. He can be reached at hallettm@ninds.nih.gov.

Paul R. McHugh, M.D., a member of the President's Council on Bioethics, is University Distinguished Service Professor of Psychiatry at Johns Hopkins University School of Medicine and professor in the Department of Mental Health at the Bloomberg School of Public Health, Johns Hopkins University. He was formerly the psychiatrist-in-chief of the Johns Hopkins Hospital. He is the author or coauthor of many books, including *The Mind Has Mountains: Reflections on Society and Psychiatry* (Johns Hopkins University Press, 2005). He can be reached at pmchugh1@jhmi.edu.

OUR FREEDOM TO CHOOSE, to make good or bad decisions about everything from which dessert to select to whether to save a life or commit a crime, seems part of our basic human nature. Long the province of philosophers and theologians, recently free will has become a question that fascinates neuroscientists. Looking for its basis in the brain has led some to argue that free will is only an illusion—a perception not congruent with the unconscious biochemical processes that they see as leading to thought and action. Two senior scientists, a neurologist and a psychiatrist, debate the meaning of free will and whether brain science can, now or ever, fully explain it. Each scientist first wrote a position statement; they then exchanged statements to write rejoinders.

Free Will: An Illusory Driver of Behavior
Mark Hallett's opening statement

My position is that free will is only a perception—our interpretation of how we experience our actions in the world. No evidence can be found for the common view that it is a function of our brains that causes behavior. I will make my argument based on research about making "voluntary" movements for two reasons. First, I am a neurologist, specifically a motor physiologist. Second, movements are easily measured. While other, more complex decisions, such as what I choose for dinner, also can be viewed as influenced by free will, I suspect that they will turn out to be analogous to movement. Anyway, such decisions often eventually manifest in movement of some kind, perhaps reaching for the cookbook or a take-out menu.

I do not doubt that I feel strongly that I have freedom of choice. And I suspect most humans have the same feeling as I do, even though I can't assess this directly. But, of course, this feeling of free will is the case only when I think about it, since most of the time I just go about my business, more or less on automatic pilot. My feeling that I have free will is a subjective perception, an element of my consciousness that philosophers call a "quale." We do not understand the biological nature of consciousness or how awareness is generated, so it is difficult to understand the physiology of any quale, including the perception of free choice. But we do know that our sense of the world is a product of our brain and that a one-to-one match between reality and that interpretation does not exist. Our introspection, our sense of what our brain is doing—while clearly useful to us and also valuable as an object of study—can be deceptive.

Looking for Free Will in the Brain

We are constantly making movements. While we certainly think we choose them freely, do we really? What would it mean if we did? The physiology of movement has been the object of intense study by scientists, and we now know the drivers of movement. These drivers include sensory input from the external world, our emotions, our biological drive for homeostasis—for balance of our physiological systems—and our past experience, including rewards and punishments that resulted from previous actions. Do these fully determine our choice or can we identify another factor, which we call free will?

The answers to these questions are easy only for the dualist, who believes in a mind separate from the brain and who thinks that free will comes from the mind. No evidence for this position can be found, however, and therefore most scientists reject it. The mind (consciousness) is a product of the brain, so if free will can be a driver of movement, we have to be able to find it in the brain. All the tools of modern neuroscience provide ways of studying this question.

Looking for free will in the brain not only is interesting for its own sake, but is also important for understanding a number of neurological and psychiatric conditions.[1] We can observe in patients with certain disorders that a relationship between movement genesis and a sense of volition is not mandatory. For example, people with Tourette syndrome often say that they cannot not act out their tics. With psychogenic movement disorders—also called conversion disorders or the old term, hysteria—the movements look voluntary, but patients say they are involuntary. In schizophrenia, movements also may look normal, but patients might say that these movements are controlled by external agents. In early Huntington's disease, the apparently involuntary chorea (rapid jerky movement) is sometimes interpreted as being voluntary. And in anosognosia (a condition in which a person who suffers disability due to brain injury seems unaware of an impairment), patients may think that they have made a movement when they have not.

Do We Freely Choose to Move?

In general, scientists need to study what we call "simplified preparations," in which it is possible to control all the variables in a situation. One such experimental situation is making a single movement of the hand or finger. People can be asked to move whenever they want to; the commonsense view is that a person consciously decides to make a movement and then makes it. Free choice has preceded the movement.

This was put to the test first in the classic experiments in the 1970s by Benjamin Libet, Ph.D., and colleagues at the University of California–San Francisco.[2] In these experiments, study participants sat in front of a clocklike timer, with a ball moving around the periphery once every three seconds. They were told to perform a simple motor task, such as flexing a finger or wrist, whenever they wanted to and, afterward, tell the investigator where the ball was (that is, what time it was) when they decided to move. An electroencephalogram (EEG) recorded their brain activity, while electromyography (EMG) recorded the electrical activity of their muscles.

What Libet learned was that participants in the study had the first conscious sense of willing the movement (called "time W") about 300 milliseconds before the onset of muscle activity (as measured by the EMG). But the EEG showed brain activity in the motor cortex beginning about 1,000 milliseconds before movement could be measured—in other words, earlier than conscious awareness of the intention to move. This early brain activity measured by the EEG probably takes place in the supplementary motor area and premotor cortex in preparation for initiating movement. This experimental result, which has been duplicated many times, appears to indicate that the movement begins unconsciously.

This does not jibe with our ordinary sense of how we operate.

Can the data from the Libet experiment be interpreted in other ways? Libet himself argued that we still have free choice, but it is confined to the ability to veto actually making a movement after the intention to move becomes conscious. This is not a strong argument, however, since the veto could also be initiated subconsciously long before the act.

Another whole set of issues revolves around the problem of the timing of subjective events. Consciousness can be deceptive, so is it possible that our sense of W, of willing a movement, is incorrect in regard to when it actually happened in the brain? A number of experiments have explored this. The results show that, first, W is not strongly linked to the time of movement onset, so whatever is going on in the brain at time W cannot be responsible for movement genesis.[3] Moreover, the brain event of W may even be later than we subjectively report. This should not be a complete surprise since humans "live in the past"—certainly perception of a real-world event has to be subsequent to its actual occurrence, since it takes time (albeit very little time) for the brain to process sensory information about the event. A recent experiment showed that it was possible to manipulate the conscious awareness of willing a movement by delivering a transcranial magnetic stimulus to the area of the brain just in front of the supplementary motor area after the movement had already occurred.[4] This suggests that the brain events of W may occur even after the movement.

If free will does not generate movement, what does? Movement generation seems to come largely from the primary motor cortex, and its input comes primarily from premotor cortices, parts of the frontal lobe just in front of the primary motor cortex. The premotor cortices receive input from most of the brain, especially the sensory cortices (which process information from our senses), limbic cortices (the emotional part of the brain), and the prefrontal cortex (which handles many cognitive processes). If the inputs from various neurons "compete," eventually one input wins, leading to a final behavior. For example, take the case of saccadic eye movements, quick target-directed eye movements. Adding even a small amount of electrical stimulation in different small brain areas can lead to a monkey's making eye movements in a different direction than might have been expected on the basis of simultaneous visual cues.[5] In general, the more we know about the various influences on the motor cortex, the better we can predict what a person will do.

In humans, various areas of the frontal cortex appear to be more active when movements are made with free choice as opposed to being triggered by sensory stimuli. A number of experiments have used neuroimaging to evaluate movements made at freely chosen times, as opposed to times indicated by sensory stimuli (for example, a sound or light). Other experiments have compared movements chosen freely from a menu, contrasted with a specifically instructed movement. These areas of the frontal cortex need more study, but their operation does not seem different in principle from that of the primary motor cortex.

From Movement to Thinking to Consciousness

The initiation of simple movement appears to be good for experimental analysis of free will. As I have described, we can understand the physiology of this process without invoking unknown forces, and our sense of volition is deceptively placed by our consciousness just before the movement occurs. This certainly gives rise to the feeling that we are freely choosing to move. But could this experimental paradigm be misleading?

Some have argued that this situation is artificial, that everything is really controlled by the experimenter, and that the free choice occurs earlier. So could it be that while the proximate initiation of movement is actually unconscious, the real free will can be found earlier? Could it be manifest in the thinking that we do before getting to the movement situation? For example, we decide, when sitting at the dinner table, that we want to eat rather than talk, and thus our hand "automatically" brings the fork to our mouth.

We don't know much about thinking, so more research is needed. But thinking will likely not be much different in principle from moving. That is to say, I expect that what happens in our brain when we initiate and process a thought will likely be similar to what happens with a movement. And, up to now, no experiment involving "free will" has identified a factor other than the routine operation of different parts of the brain.

A more detailed understanding of free will is stymied by our not understanding consciousness. Consciousness is called the "hard problem" by philosophers; it is so difficult that we do not even have a full vocabulary with which to talk about it. The best we can say now is that consciousness is awareness, and this awareness appears to be composed largely of perceptions and "qualia." Only if further research shows that free will is also a force that drives behavior can it actually be what we all naively think it is. While scientists must remain open-minded, since we have no evidence for a force of free will at the present time, I must take the position that free will exists only as a perception.

Endorsing the Obvious
Paul R. McHugh's opening statement

The editors of *Cerebrum* invited me to consider a philosophical issue—specifically that feature of the human psyche to which the expression "free will" is attached. I took them to mean that sense of responsibility felt during active choices and decisions by people aware of how, through these actions, they affect themselves and their world. The editors asked whether this "felt experience" rests upon a realistic, authentic judgment or whether this feeling of freedom is an illusion or misconception of people who are unaware that hidden factors drive them toward one choice rather than another and so render "freedom" nugatory and inoperative.

When I say this is a philosophical issue, I mean free will is a problem that philosophers like to debate. Ancient and modern philosophers (and not a few undergraduate amateurs) have offered opinions and provided reasons for affirming, denying, or qualifying the existence of freedom in the sense that we mean here. It remains a philosophical matter even though in the contemporary era neuroscientists and neuropsychologists have weighed in on the issue—mostly, and for me dishearteningly, tending to claim that free will is one of the many illusions about mental life that advancing science will put aside, much as science put paid to the "flat earth" and the "rising sun." Hence this invitation to me to present an opinion on the issue.

Why Ask a Psychiatrist?

Two questions occurred to me when deciding (freely, I presume) to accept the editors' challenge.

First: the project seemed rhetorically backward. Should I be carrying the burden of proof here? The conscious mind and all the mental experiences tied to freedom of the will—choosing, deciding, hoping, deliberating, fearing, and cooperating with others—seem as self-evident as the five senses. No one asks us to prove them "real," especially before hearing evidence that would claim they are not. Here, though, editors rule and authors follow.

Second: why ask a psychiatrist rather than a philosopher to tackle the issue? Neuroscientists challenging the concept of freedom may need more help than they realize, given the limited tools they have at hand. But, on reflection, perhaps psychiatric work may sit closer to that of neuroscientists and thus a psychiatrist may grasp their claims more sympathetically than a philosopher, even while offering evidence of a kind that, speaking to their experience, they would be less likely to ignore.

To this point, psychiatrists work with patients who fall into two great families, and the distinctions between those families do reveal matters relevant to the issue of free will. The first group encompasses all those patients whose mental difficulties derive directly from material causes (genes, infections, traumas, and so on) that change their brains. These are the patients with Alzheimer's disease, Huntington's disease, schizophrenia, and the like. In every respect, psychiatrists think about these patients in the same way neurologists think about patients with epilepsy, cardiologists about patients with heart failure, gastroenterologists about patients with peptic ulcers. They see the patients as object/organisms afflicted by some physical disruption of their bodies, brains, and minds.

Their pathologies constrain their freedom of thought and action in ways that can be put right if and when that pathology is "cured."

The other family of psychiatric patients encompasses those who provide "reasons" for their mental distress. They work under ill-fated assumptions that through "reasoning" lead them to regrettable goals. This family includes the anorexics, the dependent, and the demoralized. Although not denying that the brain is involved, their psychiatrists do not regard them as object/organisms afflicted by a pathological change in the brain but as subject/agents responding to the distressing differences between what they want and what life delivers. Their freedom is not lost but misused, and treatments customarily amount to a subtle (and occasionally not so subtle) conflict of wills between them and physicians who are striving to persuade and guide them to better ways of action by such methods as cognitive behavioral therapy.

These latter patients provide the first challenge that psychiatrists offer to those who would hold free will to be an illusion. Come see these patients where choices are the problem and where they defend their choices with arguments that frustrate their recovery. Freedom they have in abundance; it's wisdom they lack.

From experiences with both sets of patients, I concluded that freedom of the will exists in just the way consciousness itself can exist—powerful or weak, present or absent. It is far from an illusion but a basic fact of human nature on which many other facts and judgments depend.

Tying Mind to Brain

Certain philosophers and some of their neuroscience students reject this naturalistic defense of freedom by noting that it rests on descriptions of mental experiences. But, say they, brain material produces all mental phenomena, including consciousness and its expressions. These psychological "effects" emerge from the complex, mechanistic, causal apparatus that is the brain. Therefore, like all "effects" in nature, they must be lawfully determined by their "causes." What people do—and believe they choose to do—is inexorably determined by brain conditions present and past. A new predestination is born, as all psychological freedom, sensed or supposed, is illusionary.

On hearing this argument (and some form of it is far from original or even contemporary, given that Spinoza argued in similar ways), I'm struck by how it is built on presumptions rather than on a body of fact or neuroscientific discovery. Its champions presume that eventually, given the steady progress that they anticipate, neuroscientists will know all about the brain and see how mental productions are like all other productions of the body. This argument dismisses as simply a matter of the moment our present vast ignorance of just how the brain relates to mental phenomena.

But compare the difference between what we know about the way the kidney produces urine and what we know of the brain's role in producing conscious mental life. This exercise reveals that it's not just our ignorance of facts that inhibits the latter account, it's the lack of any conception of how mechanisms and products relate in it.

With urine, we have a clear grasp of how the anatomical construction and physiological actions within the nephron produce and permit a sequence of glomerular filtration of water from blood and selective reabsorbing of water and solutes in the tubule. We have an equally clear grasp of the physical characteristics of water as an ideal solvent, capable of carrying minerals and other substances through these mechanisms so as to clear the body of them. We even know, because of the work of Nobelist Peter Agre, the molecular structure and genetic construction of the aquapore in membranes that permits water to pass through them. What is mechanism, what is pathway, what is medium, and how they all interrelate for life's benefit are clear.

We know nothing of a similar kind tying brain to mind or mind to brain. We have no idea of how neural systems—distributed, modular, and parallel as various ones are—can generate anything of a subjective nature such as any element of consciousness or consciousness itself. Likewise, we have no grasp of any aspects of the elements underlying consciousness that can through their nature be tied to the anatomy or physiology of neurons, receptors, or brain systems. Much is correlated, particularly as brain imaging technology has grown, but nothing is explained.

And therefore hoping that what works in describing how organs like the kidney produce urine will work to explain the relation of brain to mind amounts to fantasy. It does epitomize the propensity of investigators to work happily away within the idiom of their successes, while ignoring issues that may demand reasoning with new idioms carrying new implications.

Scientists rightly claim that the brain is necessary for consciousness and that manipulations of the brain affect consciousness—witness anesthesia. But correlating brain events and conscious events does not explain consciousness and certainly not the vital, first-person (that is, "my") experience of consciousness on which choice and freedom rest.

As far as anyone can tell, mental "reasons" (and decisions from among them) are real aspects of nature brought into being by consciousness and, as claimed by their subjects, "free" in principle. I look for the time when neuroscientists will turn to explain how the material world can evoke these wonderful characteristics of human beings and will abandon any thought of them as illusionary.

For Freedom, Not Predestination

The answers to the philosophical question of freedom lie in the subjective realm of human life, beyond contemporary scientific capacities to explain or predict. Subjects are not the same as objects. Reasons are not the same as physical causes. Freedom is an expression of reasoning by subjects who realize that their choices determine what they make of themselves and are ready to accept the responsibility for what they fashion. To think otherwise is to give oneself over to predestination.

Ultimately in a choice between freedom and predestination, I'm for freedom and the proposition it entails—that we're responsible for making the world what it is. Indeed, this proposition organizes and justifies the therapies that psychiatrists direct.

All forms of predestination tend to disregard—and some practically disavow—the world of subjects. They shroud personal responsibility beneath foreordaining mechanisms, as Edmund in King Lear pointed out in his scornful but poetic rejection of astrology:

This is the excellent foppery of the world, that when we are sick in fortune, often the surfeits of our own behavior, we make guilty of our disasters the sun, the moon, and stars; as if we were villains on necessity; fools by heavenly compulsion; knaves, thieves, and treachers by spherical predominance; drunkards, liars, and adulterers by an enforced obedience of planetary influence; and all that we are evil in, by a divine thrusting on. An admirable evasion of whoremaster man, to lay his goatish disposition on the charge of a star.

—*KING LEAR* I, II, 128–139

I believe that we've hardly begun to grasp just what the existence of mind and freedom means. They certainly are products of the material world, but their existence tells us that material "stuff" counts in ways we've tended to ignore. There is a real presence of freedom in material, given that flesh and spirit exist together within the mystery of brain/mind. This is the deepest mystery of our kind. It is not to be denied.

Rebuttal to Paul McHugh
Mark Hallett

Obvious. McHugh calls his understanding of free will "Endorsing the Obvious." It does seem "obvious" to me that the earth is flat and that the sun goes around the earth. Yet I don't believe these things. We have sufficient scientific information that gives better explanations, and I believe those. It is also "obvious" that something cannot be a particle and a wave at the same time, but—even though I can't say that I understand it completely—I still believe that light can be so. I agree that it is "obvious" that we have freedom of choice. It is a common, if not completely universal, phenomenon, and I would even like it to be so. However, all "obvious" facts need to have scientific examination to see if they are really true.

What can we say about the human "obvious meter"? The human brain is very good, even incredible, at many things. Yet it is subject to illusion. The information from introspection is interesting and important, yet must be taken with a grain of salt. So, if we want to understand the situation, if we want to verify our obvious impression, we need some objective information.

McHugh challenges us with the anorexic patient. He thinks such patients have all the freedom of choice that they need; they are just lacking in wisdom, making the wrong choices. Psychiatrists tell them what to do, but the advice falls on deaf ears. If only they were smarter. I do agree that the "talking cure" can be effective; ideas can certainly influence behavior. However, in my view it is likely that anorexia is not solely a matter of misused freedom. Taking the medical model of disease (that something is also the matter with their brains, similar to the schizophrenic) might also be helpful with these patients.

McHugh agrees that the mind is a product of the brain. This is important, because, as I noted before, no basis for discussion can be found with anyone who believes that mind and brain are separate. So, what is the consequence? It seems clear that we need to study the brain to understand "mind" phenomena such as consciousness and free will.

Are we ready to understand scientifically all the aspects of "free will"? I am not so sure, and in this McHugh and I do agree. We must confront a big barrier, the understanding of consciousness itself. However, I believe that we can make a start on the issue of free will, and it may well be that such a start will be helpful in understanding consciousness as a whole.

Philosophers can be useful in the search for answers. Indeed, we need all the help we can get from any serious line of investigation about the nature of the physical world. Physicists, for another example, should also be helpful. The topic of free will is not solely a philosophical issue anymore. It is possible to gather objective data about it. Moreover, it is important to do so, because the consequences are moral, social, medical, and legal. Let's take our heads out of the sand, and be willing to accept the facts as they are exposed. Even if it does not make for good poetry.

Rebuttal to Mark Hallett
Paul McHugh

Neurologists and psychiatrists collaborate as specialists studying and treating human disorders arising from the brain. But like competitive siblings, they tend to vie for attention, particularly over who has the better—more scientific, enlightened, "grown-up"—approach to ultimate questions of human mind and nature.

Both work with patients and thus may not know as much about the healthy mind as they think! But psychiatrists attending to the way patients behave ask "why?" whereas neurologists attending to the way patients move ask "how?" These questions represent the "top-down" as against the "bottom-up" approach to the mind/brain mystery. The answers inform and enhance the study of human activity, but they bring about differing professional emphases and distinct clinical experiences that probably explain the differing notions on free will revealed in this friendly epistolary debate.

No psychiatrist, though, would think that the experiment of Benjamin Libet that Hallett mentions constitutes a test of human free will. That experiment—seen as a meritorious "simplified preparation" by Hallett— would seem to most of us psychiatrists as simpleminded in every sense of that term. It misses the point at issue by assuming that all "voluntary" motor actions deserve to be considered examples of free will.

Hallett describes what the experimental subjects were told to do: They were to "[flex] a finger or wrist, whenever they wanted to." But this simply translates into "move when the impulse grabs you." Given these instructions, surely no one should be surprised that neurophysiologic techniques can pick up traces of the emerging and unopposed impulse before it strikes the subjects' consciousness. What else would one expect when studying a capricious act with neither consequence nor significance to the actor?

Free will surfaces with deliberation, not impulse. And, by definition, deliberation lies in consciousness, where matters riding on a decision are grasped. The experience of deliberation is a perception (as Hallett notes), but it's a complex perception of implications tied to choices and responsibility.

Bottom-up neuroscientists find consciousness a "hard problem," as my colleague also notes. But nonetheless, consciousness is where the action is with free will. What to identify as a function of consciousness divides us in this debate.

I, thinking "top down," hold that with consciousness nature brings freedom into animal life. And nature has done so progressively and evolutionarily. Individuals survive and flourish when they can pull together a presenting situation and past experience under the light of consciousness before deciding what to do.

Humans carry this natural freedom beyond issues of individual survival. Human reflections and the choices they engender encompass judgments that sustain communities—deliberating as humans and no other creatures do on matters of right and wrong, good and bad, "me" and "not-me." From these perceptions and the actions they evoke, personal integrity, cultural reliability, and truth that matters grow exponentially.

Percy Bysshe Shelley asserted, correctly I believe, that "poets are the unacknowledged legislators of the world." Poets can play that role because they are the most conscious of our kind, and, as the ultimate top-down thinkers, they identify how the freedom embedded in our consciousness transforms us, gives dignity to our fellow man, and charges us with responsibilities to choose the better courses of action in our lives. In deliberating over what we can notice and how we should respond, poets exhibit more open-mindedness than many neuroscientists. These latter folk, laboring under a metaphysical obsession with dualism that hampers their bottom-up reasoning, can miss—or, as here, dismiss—what we all know about our mental experience and its yield. Indeed, neuroscientists sometimes take up positions, strike postures, and tout experiments that sell us humans short—all of us, themselves included.

CHAPTER 10

Stress and Immunity

From Starving Cavemen to Stressed-Out Scientists

by Fabienne Mackay, Ph.D.

Fabienne Mackay, Ph.D., is director of the Autoimmunity Research Unit at the Garvan Institute of Medical Research in Australia, and an adjunct professor of medicine at the University of New South Wales and the University of Sydney. A molecular biologist and immunologist, she studies autoimmune diseases and the interaction of the immune and nervous systems. She can be reached at f.mackay@garvan.org.au.

Scientists in Australia have recently discovered the first clear molecular process that helps to explain how stress suppresses our immune defenses and makes us more vulnerable to getting sick. The author describes how the brain and the immune system talk with each other through a tiny protein called neuropeptide Y, which plays a surprising dual role in how our bodies deal with stress. Has a biological system that worked well for early humans faced with starvation turned against those of us living with the many new stresses of modern society?

HAVE YOU EVER NOTICED that during periods of intense stress you are more prone to catch a cold or to come down with the flu? Scientists have long known that psychological stress, whether from a sudden trauma or a routine event of daily life such as a difficult commute, adversely affects our immune responses. How can we explain this? What are the cellular and molecular mechanisms that lead to a compromised immune system during stressful times?

We now know that many of the body's important systems are closely interconnected. For instance, the immune system has connections to the nervous system and to the metabolic system. Many molecules that are used by the nervous system are also used by immune cells, and molecules in one system can have an important, and different, effect on the other. Now for the first time, interdisciplinary studies of a molecule found in both the nervous system and the immune system have revealed a clear link between stress and immune suppression.

Stress and Its Consequences

Psychological stress may be defined as any external condition or trauma that disturbs an individual's psychological and physical well-being. This stress is subjective, as situations that stress one person may not stress another (for instance, I find air travel particularly unpleasant).

People's individual genetic makeup, combined with their life experiences, especially when they are young, results in a wide spectrum of responses to psychological stress. Stress is a worldwide challenge to health. Sometimes it comes from extreme conditions of poverty, starvation, persecution, or war. It can also be the result of caring for a sick family member, the loss of a loved one, troubled relationships, being in an occupation that involves a high level of responsibility or danger (police, airline pilots, air traffic controllers, firefighters), or the heavy workloads and the challenges of balancing professional and family life that are common in the Western world.

The most often noted manifestations of psychological stress are mental and physical fatigue, anxiety, anger, and depression. The second-most-common symptoms are immune disorders, especially an increased susceptibility to viral infections (herpes, flu, colds), bacterial or fungal infections (pneumonia, mycosis, meningitis), and inflammation (stomach ulcers, gastritis). Stress can also trigger allergies, asthma, autoimmune diseases such as juvenile diabetes, and inflammatory bowel disease. A link between chronic stress and the inability of immune cells to mobilize a strong defense against cancer has also been suggested.

Perhaps as a result of the considerable sources of psychological stress in our lives, the "wellness" industry has emerged, providing relaxation techniques, yoga, therapeutic massage, and psychological counseling. The irony is that for many people, adding stress-relieving activities to their busy schedule often creates additional stress. Moreover, our genes play an important role in how we react to stress, and so any positive response to relaxing activities varies considerably from person to person. The reality is that while relaxation cannot hurt, a sea change is not always possible, and a change of genetic makeup is impossible. Therefore, science must find new solutions to alleviate both the psychological and the immune-system side effects of stress. Here, then, is an opportunity for immunologists and brain researchers to collaborate.

The Many Roles of Neuropeptide Y

A considerable amount of evidence now supports the existence of "crosstalk" between the immune system and the nervous system. The nervous system, including the brain and peripheral nerves, can stimulate or inhibit various parts of the immune system. Conversely, the immune system can influence functions of the nervous system through the release into the blood of factors such as "messenger" proteins called cytokines, which are produced by white blood cells.

Interestingly, a number of peptides (very small proteins) with neural or neuroendocrine functions also have potent antimicrobial activities. One possibility is that the nervous system uses these peptides as a defense against infection at nerve terminations throughout the body. New research is also showing that one of these peptides, neuropeptide Y (NPY), appears to have other functions in the immune system.

NPY is a hormone that is secreted under stressful conditions by cells in the brain, and possibly by some immune cells, and is then carried throughout the body in the blood. The NPY hormone uses at least five receptors located on the surface of other cells, such as certain immune cells, to attach to those cells. These receptors are called Y1, Y2, Y4, Y5, and Y6.[1] NPY's effects are as varied as its receptors. A range of research studies has shown that this hormone is involved in functions ranging from eating behaviors to anxiety, memory, seizures, pain, drug addiction, circadian rhythms, cardiovascular disease, bone mass development, and blood vessel dysfunction. Because so many NPY receptors can be found, and NPY can carry out so many different functions depending on the receptor to which it binds, researchers have had trouble understanding what function is controlled by what receptor. To study this, scientists have genetically engineered animals, such as mice, that lack one or more of the NPY receptors. These new animal models have been instrumental in understanding the respective contributions of each NPY receptor in behavior, cardiovascular function, and metabolism, but little research on the immune system had been done until very recently.

A No-Brainer Initiative

Like most immunologists, I have focused my interest strictly on the immune system and not on the brain. It took some rather unusual circumstances to change the way I now look at immune defenses. At the Garvan Institute, my immunology research group works side by side with researchers from different disciplines. For a long time, this structure has been viewed as a handicap, and many scientists have campaigned for a more specialized structure with a critical mass of scientists focused on the same area of research, rather than being spread thin with smaller groups that have little in common. Like members of most institutes, Garvan scientists meeting in a corridor talk about their busy schedules and the pressure to find more research funding and to publish, lamenting that the stress will kill them. One day they finally asked the relevant question: why will the stress eventually kill them? That question precipitated my collaboration with Herbert Herzog, Ph.D.

Herzog, who directs the neuroscience program at the Garvan Institute, is an expert in NPY in the nervous system and the biology of stress. He has developed genetically modified mice that lack NPY or one or more of its numerous receptors, but he knows little about the immune system. Because indirect evidence from earlier research suggested that NPY had a role in the immune system, Herzog and I decided to embark on a project that would study the immune system of mice that were deficient in NPY or its Y1 receptor, chosen for its expression pattern in the immune system. He bred special mice for our research that had a genetic background suitable for immunological studies and gave us a background in NPY biochemistry.

Julie Wheway, a talented Ph.D. student in my lab at the time, started the pilot experiments with NPY and soon came back to my office with data clearly showing immune disorders in mice that had been injected with NPY and in mice lacking the Y1 receptor. Our further research yielded evidence that NPY, working through the Y1 receptor, suppresses the action of key immune cells named T cells. T cells and other adaptive immune cells are part of the body's defense. They help us fight specific infections and also participate in surveillance for cancerous cells.

T cells are activated by another set of immune cells that are the body's first line of defense. These are called innate cells and are mostly macrophages and dendritic cells. These innate immune cells patrol the human body in search of microbes or cancerous cells, which they kill and consume. The dead invaders are placed on the surface of the innate immune cells like small "hunting trophies." The patrolling cells then report back to the nearby lymph node, where they inform T cells that they have found an infectious problem somewhere, evidence of which is displayed on their cell surface. Upon receiving this warning, T cells are activated for immediate response, dividing quickly and forming armies of cells ready to go help to clear the infection. T cells also help another type of adaptive immune cell, named B cells, which produce antibodies, small molecules that bind to microbes and help clear them.

Our discovery that immune T cells can be suppressed by NPY was soon followed by another discovery. Wheway and I found that, in contrast to NPY's suppression of T cells, some NPY is necessary for the patrolling of innate immune cells to be fully functional. However, too much NPY, such as that produced during chronic stress, may create overzealous patrolling of innate immune cells and, consequently, inappropriate inflammation in tissues. So you can imagine how, in the same overstressed person, a doctor might detect inflammation caused by macrophages or dendritic cells that have been overactivated by NPY and, at the same time, find a persistent viral infection that has flourished because T cells have been suppressed by NPY.

What Was Mother Nature Thinking?

This research shows that NPY and its receptor Y1 have a critical but dual role. Why might Mother Nature have designed this "yin-yang" way of regulating the immune system?

We still don't fully understand the natural advantages of this dual mechanism, and it will be important to integrate this effect within a general understanding of how other brain factors regulate immune function. But let's go back in time through human evolution and try to better understand what the main purpose of NPY could have been when the causes of psychological stress were very different than they are now.

Early in the human race, starvation was a main source of stress. NPY in our ancestors played a critical role in adapting to scarce food supplies. Scientists studying the neuroendocrine system have learned that NPY produced in the brain stimulates food intake, lowers energy expenditure, and facilitates the storage of fat.[2] Research by Herzog showed that part of the function of NPY is also to inhibit the energy-consuming function of bone formation, as bone density is increased in mice lacking the Y2 receptor for NPY.[3]

In addition, raising armies of T cells through cell-division proliferation is highly energy-consuming. Perhaps NPY secretion during stress is meant to increase the activity of patrolling innate immune cells to enable them to deal with infectious invaders for the short term, while inhibiting T cell activation to save energy. In general, it seems that the NPY system has been designed to help mammals living in a primitive environment cope with harsh times—starvation—by allowing fat storage, while reducing the expenditure of energy in various parts of the body, including the immune system. Therefore, NPY might have originally been a fundamental survival mechanism that allowed mankind to survive on lower supplies of body energy.

Starvation remains a reality for most biological systems, including many human populations. However, it seems that the function of NPY and its receptors, originally designed to help us survive harsh times, has been diverted from its original purpose by the novel forms of stress associated with life in today's world. Mother Nature could not have predicted factors of modern life such as industrialization, technological progress, hygiene, and food abundance, which, while bringing undeniable benefits, also change the environment in which our bodies function. Is it possible that stress-induced NPY, once designed to help humans survive in lean times, is working against us and making us sick?

Before we get too depressed about this possibility, let me make a key point. In our society, where people say that they get stressed about threats from terrorists, nuclear weapons, climate change, overpopulation, pollution, the quality of food, new epidemics such as avian flu, cancer, obesity, and cardiovascular disease, one must relax a minute and consider that the life expectancy of humans has never been higher. Modern-day stress can indeed make us sick, but modern-day medicine does a good job of keeping us alive longer and longer. Cavemen may have enjoyed clean air, pure water, plenty of exercise, and free-range organic meat, yet few of them survived past thirty years of age. The occasional NPY-induced shutdown of their T cell function to save body energy had serious limitations.

The Road Ahead

Our discovery of NPY's critical role in the immune system was both totally unexpected and very powerful. It provides the first clear molecular mechanism to explain in great part how stress suppresses our immune defenses and makes us sick. This understanding is changing the way scientists look at the immune system and suggests powerful new possibilities for treating immune diseases at the interface of the brain and the immune system.

We now know that the brain talks to the immune system, but it is also likely that the immune system talks back to the brain. The challenge is to decode the exact "language" that these two biological systems are using to communicate. No doubt NPY and its receptors will provide additional exciting clues to the brain's regulation of normal immune functions and also their dysregulation in immune diseases.

A last important point is that we owe our discovery of neuropeptide Y's versatility to the willingness of researchers in two disciplines to work collaboratively. It should happen much more than it does, but getting scientists from one discipline to work with those in another discipline has not always been easy. Often they don't talk the same language, because the cells and molecules that they study are usually different, and they consider the particular biological systems that they study as operating independently from other systems. But, now, for example, new technologies allow us to identify genes normally found in one system that are also associated with cells in a completely different one. The key is not for a scientist to attempt to be an expert in two separate fields, such as immunology and brain research, but rather to bring together experts into a scientific framework that fosters interdisciplinary approaches.

Scientists (particularly immunologists) often detach themselves too much from the outside world, concentrating all their energy and attention on a specific field. Yet an inquisitive scientific mind cannot help but question the invariable outbreak of flu, bronchitis, and common colds among stressed colleagues immediately prior to an important grant submission deadline that will decide whether their research, and possibly their salaries, will be funded or not. The observation that psychological stress often leads to immunosuppression has always been part of our lives; indeed, many studies have clearly and statistically demonstrated this important relationship. Until we design still more interdisciplinary ways of working to lead us to new discoveries, I guess we can still count on getting sick to take us where we need to go.

The observation that changes in our way of life have resulted in a deviation of NPY biology from its original function is not something to be stressed out about but something that we need to understand. As an immunologist, I am gratified that my lab discovered that NPY and its Y1 receptor play a key and previously unappreciated role in the immune system.[4] I know that this is only the beginning and that other NPY receptors are also important in the immune system. While my plans to continue to decrypt this system may sometimes seem overwhelming (a last stressful thought), important new discoveries are just around the corner. Among the potential advances from this work may be new therapeutic solutions, new biological concepts, and a new look at life with stress-induced NPY.

CHAPTER 11

Harnessing the Brain's Power to Adapt After Injury

by Michael E. Selzer, M.D., Ph.D.

Michael E. Selzer, M.D., Ph.D., is a professor in the Department of Neurology and director of the Center for Experimental Neurorehabilitation Training at the University of Pennsylvania School of Medicine, as well as director of Rehabilitation Research and Development in the Department of Veterans Affairs. His research focuses on spinal cord regeneration and translating basic mechanisms of plasticity into rehabilitative therapies for central nervous system injuries. He is the lead editor of the *Textbook of Neural Repair and Rehabilitation*. He can be reached at michael.selzer@uphs.upenn.edu.

Whether a brain or spinal cord injury is caused by a weapon of war, an acci-dent, or a disease such as stroke, rehabilitation focuses on enabling people to make the most of what functions they still have. Physical, occupational, and speech therapy, counseling, and education can go only so far, however. For neurorehabilitation to offer the hope of curing the underlying brain damage, writes an expert in the field, it must look to basic science and better clinical trials to put to work the power of the brain's plasticity.

MORE THAN 300,000 AMERICANS EVERY YEAR receive a head injury severe enough to require medical attention, and about 75,000 of them end up with permanent neurological damage. Many more people suffer brain damage from stroke, multiple sclerosis, or Alzheimer's and other diseases. The wars in Iraq and Afghanistan have focused the public's attention on the potentially devastating nature of brain injuries because soldiers, whose bodies are protected by effective armor, are able to survive gunshot attacks and percussion injuries from IEDs (improvised explosive devices). Their hearts, lungs, and other vital organs are protected, and within minutes of an attack their vital body functions receive medical attention in the field. As a result, brain trauma represents a larger proportion of nonfatal war injuries than was seen in previous conflicts.

Men and women come home from the war with functional impairments that range from severe paralysis to problems with memory or speech to the more subtle but often devastating loss of the mental abilities that enable a person to concentrate on a problem and have the initiative to solve it. Without the ability to handle complex information, organize it into meaningful categories, set reasonable goals, and make good decisions, they cannot resume their careers or even successfully navigate daily life.

To give hope to our veterans suffering from traumatic brain injury as well as millions of people worldwide who have some kind of brain damage, we must seek to reverse this damage, not just ameliorate it. The current treatments for traumatic brain injury are inadequate. Rehabilitation medicine relies almost exclusively on techniques aimed at making the most of what neurological function is left, not changing the deficits themselves. Although in recent years, these techniques have received more careful scientific study, many in the field have sensed that progress has reached a standstill. For this to change, we must take an approach traditionally considered outside the definition of rehabilitation medicine: focusing on the underlying neurobiology and allowing ourselves the aspiration of actually curing brain damage through harnessing the brain's power to change and heal.

The Traditional Approach

Rehabilitation medicine has been well ahead of the field of medicine in general because it emphasizes the importance of assessing patients' functioning in practical tasks and of their quality of life. Traditionally, neurorehabilitation has employed physical, occupational, and speech therapies, evaluated the home and work environments, and worked hard to educate patients and families on strategies to compensate for lost functions. But as important as these strategies are, they cannot fully restore normal function, and research into improving their design and application does not appear to be leading to profound advances.

Typically, we retrain a patient to accomplish a specific task, using practice and repetition, but that one task is as far as it goes—what has been learned does not transfer to other tasks. For example, a stroke in the speech area of the left cerebral hemisphere may result in impaired language (aphasia), which includes difficulty finding the correct names for objects. A person may recognize a watch and understand its purpose but be unable to come up with the word "watch" when asked to identify one. With extensive practice, the aphasic person can eventually relearn to say "watch," but this does not result in improved naming of other objects. In fact, after learning 100 words in this way, the person would take virtually the same amount of time to learn the next 100 words.

Similarly, physical therapy, including passive stretching of tightened right arm and hand muscles and lots of practice performing tasks with that arm and hand, may improve those arm and hand functions, but only practicing walking can improve a limp that resulted from the same stroke.

These examples illustrate the limitations of traditional rehabilitation approaches. The day does not hold enough hours to practice every single activity and, even if it did, improvement would be incomplete.

The Plastic Brain

Although scientists long thought otherwise, the brain has the ability to adjust the strength of its electrical connections and even to sprout new connections, in order to cause partial recovery of abilities previously lost because of an injury or stroke. Enthusiasm for the concept that the brain can adapt to injury has fluctuated, but experiments in monkeys and other animals have rekindled interest in the possibility that the brain is more changeable than was once thought. Perhaps we could harness that "plasticity" to help people recover from brain injuries.

To give an idea of what plasticity might mean, let me mention a few key experiments. In the 1980s, Michael Merzenich, Ph.D., at the University of California–San Francisco, and other researchers showed that when a small body part such as a finger is amputated in a monkey, the nerve cells in the sensory part of the cerebral cortex that used to respond electrically to stimulation of that body part did not become quiet. Instead those nerve cells began to respond to stimulation of neighboring body parts. Similarly, studies by Randy Nudo, Ph.D., at the University of Kansas showed that if a tiny stroke is produced by blocking the blood flow to a small part of a monkey's motor cortex, the part of the body that used to move in response to electrical stimulation of that area of cortex would now move when nearby areas of the brain were stimulated.

In recent years, functional imaging studies by Richard S. Frackowiak, M.D., D.Sc., in London, and many other groups around the world, have confirmed that the brain can change its responses in human stroke patients in ways similar to what was found in monkeys. This has also been shown by experiments using transcranial magnetic stimulation of the human cortex. Studies by Leo Cohen, M.D., at the National Institutes of Health (NIH), his protégé, Alvaro Pascual Leone, M.D., Ph.D. (now at Harvard), and many others have shown that recovery from stroke is accompanied by plastic changes in the way that electrical activity in the brain triggers movements in the body. These changes go beyond behavioral adaptation to injury.

The Challenges of Studying Rehabilitation Therapies

These demonstrations of inherent malleability of the human brain have triggered a period of intense experimentation seeking to harness that plasticity. But so far we have been frustrated by the limits imposed by the biology of the brain and by the difficulty in doing human experiments that clearly tease out the effects of therapy. Not only do limits exist on how much practice can be devoted to relearning cognitive and motor tasks, but it has turned out that the brain's plasticity has severe limits as well. Furthermore, it is difficult for researchers to determine how much improvement is attributable to a particular therapy, how much to what is called the placebo effect, and how much to the spontaneous partial recovery that is the norm after stroke or brain injury.

Let us look at this last point first because it affects everything else. Most spontaneous improvement occurs within the first month after a brain injury, but some additional gains occur over the course of three to six months. Unless carefully controlled studies are performed on patients with similar injuries and functional impairments, and those patients are also matched for age, sex, and other important variables, and randomly assigned to groups receiving different treatment regimens, it is not possible to say whether a given therapy has had a beneficial effect. This is especially true when the treatment is tested during the first few months of recovery.

We also know that any therapy causes a placebo effect—improvement for reasons not specific to the therapy itself but to the complex psychological and biological effects of the attention, encouragement, and expectations associated with being a part of the testing process. The placebo effect adds further variability to experimental results already made variable because of the element of spontaneous improvement. The placebo effect can be minimized, but not eliminated, by being scrupulously careful that neither the patient nor the investigator knows which treatment is the one being tested and which is the placebo—an alternate, usually inactive treatment.

Having all patients in an experiment carefully matched, randomly assigning them to groups, and controlling for the placebo effect are the key components of what is called the prospective, randomized, double-blind, placebo-controlled clinical trial (RDPCT), which has become the gold standard of medical research. Controlling for the placebo effect is relatively easy to do when the treatment is a pill, because it is possible to design placebo pills that look just like the test drug. But when the treatment is a physical therapy, how do you hide the real treatment from both the patient and the therapist? Sometimes this is impossible. But in a single-blind clinical trial, the treatment being tested can be matched with a control treatment that is of similar intensity and duration. In that case, the therapist knows which treatment is being tested, but the patient does not.

In addition to the challenges of setting up a double-blind test for a physical therapy, other aspects of rehabilitation can be at least as difficult to study in controlled clinical trials. Designing a study that can actually delineate the benefits of a particular form of education and family counseling, vocational rehabilitation, or the use of prosthetic devices for amputated limbs can be daunting.

In part because of these challenges, the professional rehabilitation community has been slow to submit its treatments to randomized controlled clinical trials, and only recently has this been done on a scale large enough to yield reliable results. Two recently completed high-profile studies illustrate the complexity of the problem. I describe them briefly here because they made so much news in the medical community and because they were the first NIH-sponsored, randomized, multi-center, controlled trials of any physical therapy. The results of both trials suggested that, with intense enough practice, patients with injuries to the central nervous system could gain some improvement, but they left the field of medicine in the dark as to what influence the precise mode of therapy had.

The first study, led by Bruce Dobkin, M.D., a neurologist and neurorehabilitation expert at the University of California, Los Angeles, was a test of whether treadmill training, which strengthens a spinal cord circuit that generates walking and running movements, would work better than conventional physical therapy in people with a spinal cord injury (the Spinal Cord Injury Locomotor Trial, or SCILT). One hundred forty-six people who had experienced an incomplete spinal cord injury were studied at six medical centers. The patients all received one hour of physical therapy per day for twelve weeks. Half of them were put in a harness that supported part of their weight while therapists helped them walk on a treadmill, and the other half received more-conventional physical therapy.

When their walking was tested later, both groups of patients had improved equally. From one point of view, this outcome was disappointing, but a very interesting feature emerged. Both groups of patients in this study, which involved more intensive physical therapy than most patients ever get, improved more than was expected for most patients with spinal cord injuries. In other words, it may be that the intensity of training was more important here than the precise mode of training.

The second clinical trial was the Extremity Constraint-Induced Movement Therapy Evaluation (EXCITE), headed by Steven L. Wolfe, Ph.D., P.T., a physical therapist at Emory University School of Medicine. Constraint-induced movement therapy (CIMT) for people who have experienced a stroke and lost function in an arm or hand involves forcing the affected arm to work by binding the unaffected arm in a mitt or sling. It can improve function of the affected arm as long as patients had some hand and wrist movement to begin with, and this improvement can persist for at least a year. The goal of the EXCITE trial was to test this benefit. Therefore, 222 stroke patients at seven academic medical centers were randomly divided into two groups. One group was treated with CIMT, in which the good arm was kept in a restraining mitt and the bad arm was engaged in an intensive regimen of repetitive task practice and behavioral shaping. The control group was treated with "usual and customary care," but the intensity and the amount of time devoted to their physical therapy were not specified.

Although, as expected, the CIMT group showed greater improvements in arm function than the control group and the benefits persisted for at least one year, it is nonetheless difficult to evaluate the results. On the one hand, the control group did not receive the same intensity of therapy as the CIMT group, which means that it is not possible to determine how much of the improvement seen in the CIMT group was a consequence of the specific mode of therapy they received and how much was related to the amount of therapy in general. On the other hand, the main purpose of the constraint therapy was to force the bad arm to be used more than it would have been if the patient were free to use the good arm for everything.

One interesting aspect of the EXCITE trial is that many of the patients who were enrolled had experienced their strokes long enough in the past that they had already achieved most of the spontaneous recovery and even the improvement as a result of physical therapy that would normally be expected. The improvement they experienced as a result of the EXCITE trial showed that intense physical therapy can result in additional benefit, even late after stroke. Nonetheless, this trial still left us with the need to do more research.

The results of both the SCILT and the EXCITE trials, reported in journal articles in 2006, reinforced two aspects of research on physical therapies. First, the therapies can deliver some improvements in function, but not cures. Second, the intensity of therapy appears to be important in the amount of improvement, but it is still not clear whether the precise technique used is important.

Rehabilitation Needs Basic Science

Just as medical rehabilitation has lagged in developing evidence-based practice, the field also has been slow to adopt a basic science underpinning, and it is tempting to conclude that the two phenomena are related. Although the central nervous system may be more plastic and capable of more adaptation to injury than was believed previously, severe limits exist to what can be accomplished by practice and exercise alone. If we are to go beyond these limits, we will have to intervene more actively in the basic biology of the nervous system. Basic science studies provide a rationale for clinical studies and expand our view of what might be possible.

Take spinal cord injury as an example. The most famous case is that of the actor Christopher Reeve, who became totally paralyzed from the neck down as a result of a horseback riding accident in 1995. Before his death in 2004, he engaged in very intensive physical therapy, including treadmill training with the aid of physical therapists. Because he had financial resources beyond those of most people with spinal cord injuries, and the determination to participate actively in his own treatment, he was able to avail himself of the most up-to-date and most intensive therapies that exist. One journal article, in fact, indicated that late in life he had begun to regain some sensations below the neck and to experience flickers of movement. But despite this improvement, he remained essentially totally paralyzed in both arms and both legs. Reeve himself was convinced of the need to apply basic science research to the problem of spinal cord injury, and he formed a research foundation dedicated to finding ways to regenerate nerve pathways in the injured spinal cord. He also lobbied intensively for stem cell research, believing that this had great potential for repairing the injured brain and spinal cord.

We have made real progress in understanding some of the mechanisms that control nerve regeneration, as well as some of the factors that explain why this is so difficult in the human central nervous system. Some animal experimentation has yielded promising results. Yet, in my opinion, we have a way to go before we can apply this knowledge successfully to human brain and spinal cord injury. Among the many important questions remaining to be answered are these:

1. Since so much of what we think we know about regeneration is derived from experiments on immature nerve cells, are the mechanisms of regeneration in the injured mature nervous system the same as those that apply to the developing embryonic nervous system?

2. Since the vast majority of experiments in regeneration of nerve pathways have been done in rats and mice, how predictive are these experiments for results in human patients? Apart from molecular differences, rodents are much smaller than we are. Nerve fibers may have to regenerate much farther in humans in order to achieve the same level of reconnection that underlies functional improvement in smaller animals.

3. Even if sufficient nerve regeneration can be achieved, will the connections made be specific enough to underlie real function?

4. How helpful are stem cells? Can they survive after transplantation into the human spinal cord or will they be rejected? Can they replace damaged neurons or will they serve only as sources of chemical substances that support survival and growth of the brain's own nerve cells?

5. Will we be able to identify a single approach that is so fundamental that it can yield dramatic improvements in recovery from brain injury, or will we need to develop a cocktail approach, using multiple treatments simultaneously?

6. Will approaches that enhance regeneration in one circumstance, for example spinal cord injury, also work in other situations, such as stroke or traumatic brain injury?

When these and other questions are answered—and I am confident that they will be—we will be able to offer a much more profound recovery than we can to our veterans returning from Iraq and Afghanistan today with brain and spinal cord injuries. But even in the shorter term, a broadening of the research agenda in neurorehabilitation to include addressing the fundamental biology underlying disabling brain disorders can help the field of rehabilitation medicine advance in ways that go beyond offering patients better functional recovery. It can help attract the best and brightest graduates of medical schools to a specialty that has, to date, specifically ruled out cure from its research agenda by defining itself as a specialty that seeks to optimize function in people with fixed anatomical and physiological impairments. Rehabilitation medicine has left the possibility of curing the impaired anatomy to other specialties, such as neurology and neurosurgery.

But now with new influences, especially from the neurorehabilitation community, a multidisciplinary collection of neurologists, neuroscientists, physical therapists, physiatrists, and other rehabilitation professionals is coming together to attack the problem of central nervous system injury. These researchers subscribe to the notion that rehabilitation can best be achieved by understanding the fundamental neuroscience that underlies neurological disabilities. Because this new generation of rehabilitation scientists is committed to intervening in the underlying pathophysiology, the field of rehabilitation medicine is undergoing a dramatic enhancement in its research mission and its attractiveness to talented medical school graduates.

Signs of Change

This new vigor is reflected in the expansion of the research supported by the Rehabilitation Research and Development Service of the Office of Research and Development in the Veterans Health Administration and by the National Center for Medical Rehabilitation Research, the two largest funding agencies for rehabilitation research. In addition to developing evidence of the effectiveness of traditional rehabilitation interventions, these agencies are supporting research in such futuristic areas as nerve regeneration in the brain and spinal cord, brain-computer interfaces for communication and motor control in paralyzed persons, robotic-assisted physical therapy, electronic-assisted prosthetic limbs, and prosthetic neural circuitry. In the area of spinal cord injury, approaches aimed at neutralizing the molecular inhibitors of nerve axon regeneration have already reached the stage of human clinical trials.

This progress is exciting because the pharmaceutical industry is now ready to gamble its own resources to test the therapies developed in animals by scientists who are supported by the National Institutes of Health, the Department of Veterans Affairs, and other public and private research organizations in the United States and abroad. In one such clinical trial by the pharmaceutical company Novartis, antibodies against a molecule (whimsically called Nogo) found in the myelin sheath that insulates axons are being injected into the spinal fluid of patients with severe spinal cord injuries to see if these patients will be able to regenerate the severed nerve connections. After a preliminary study of more than twenty-five patients, no serious toxicity was found, and the study will be expanded to more patients to gain preliminary evidence for effectiveness.

Alseres Pharmaceuticals is testing a second approach, in which an inhibitor of the enzyme RhoA is applied to the membrane covering that surrounds the spinal cord. RhoA is part of the signaling pathway that mediates the effects of Nogo and other growth-inhibiting molecules. Study of the use of this inhibitor in more than thirty patients revealed no serious toxicity, and the clinical trial will be expanded to test for efficacy. It is too soon to know whether either of these treatments will promote axon regeneration and functional recovery of human patients with spinal cord injuries, but because successful experiments in animals were reported by highly respected investigators, the clinical trials have attracted a great deal of attention. Moreover, animal experiments suggest that approaches based on neutralizing Nogo might also help in the recovery from stroke, a condition much more common than spinal cord injury.

The association of the clinical neurorehabilitation community with this and other work at the forefront of translational research has done much to increase the acceptance of rehabilitation medicine by the general medical community. This in turn has enhanced the influence of medical rehabilitation on the overall field of medicine. Clinical trials now routinely employ the types of tests of overall patient physical and mental functioning, as well as indicators of quality of life, that have been developed by the rehabilitation research community. Doctors now recognize the essential truth that improvement in a physiological function is not necessarily helpful unless it enhances people's abilities to resume their role in society and their satisfaction with their life.

Both enlarging the scope of research on rehabilitation from brain injuries to encompass basic science, particularly related to harnessing the power of plasticity, and expanding the goals of rehabilitation beyond improving function (as important as that is) to actually reversing the disabling damage are beginning to yield important benefits. The field of neurorehabilitation has added impetus to the efforts to cure stroke and spinal cord injury, increased the credibility and influence of rehabilitation among medical specialties, and extended its influence in ways that are very beneficial for the public. If we combine basic research on plasticity and regeneration with more and better clinical trials, rehabilitation medicine will no longer be limited to ameliorating symptoms but can look toward dramatic improvement and even cures, not only for the thousands of injured returning soldiers but for everyone (and that could be all of us someday) with a head injury or a neurological disease.

Disclaimer: The views expressed in this article are those of the author and not of the Department of Veterans Affairs or the government of the United States.

CHAPTER 12

"Go" and "NoGo"

Learning and the Basal Ganglia

by Michael J. Frank, Ph.D.

Michael J. Frank, Ph.D., is an assistant professor in the Program in Neuroscience, Cognition, and Neural Systems at the University of Arizona. His research focuses on understanding the neural mechanisms underlying learning, decision making, and working memory. He has also developed neural network software for this purpose. He can be reached at mfrank@u.arizona.edu.

Many human behaviors—perhaps more than we would like to think—are, in essence, reflexes programmed into our brains when we are rewarded or punished for taking a particular action. New research is showing how the basal ganglia, deep inside the brain below the cortex, are important in learning from feedback, in the formation of good and bad habits, and even in brain disorders as diverse as Parkinson's disease, ADHD, and addiction. Reflexes deserve respect, writes the author, and understanding how people differ in learning from positive or negative feedback may have implications for education as well as for treating diseases in which the basal ganglia's systems go awry.

BEFORE YOU READ ANY FURTHER, grab a glass of your favorite beverage and set it down (no drinking yet). Done? Now, reach both hands around your back and touch your pinkies together. Then quickly take a sip of your drink. Go on.

Did you do it? If so, the next time you find yourself in a similar environment you will have a greater chance of spontaneously repeating this round-the-back pinkie act. Although that possibility may sound strange, your brain is actually programmed to reinforce actions that are immediately followed by rewards. This is especially true when the reward is unexpected (you probably did not expect to have a treat when you began to read this article).

Although most of us feel like we are in control of our actions, many of those actions can also be explained by principles of learning that are embedded in our neural machinery. Of course, this machinery is inordinately intricate and complex, involving several interacting systems, each with millions of neurons, billions of connections, and multiple neurotransmitters, all evolving dynamically as a function of genes, time, past experience, and current environment. But neuroscience is shedding light on how circuits linking two parts of the brain, the basal ganglia and the frontal cortex, contribute to learning both productive and counterproductive behaviors, and even to some neurological disorders. Those circuits can, for example, help account for genetically driven individual differences in whether we learn best from positive or negative reinforcement, and understanding them provides insights into decision making in people with Parkinson's disease, attention-deficit/hyperactivity disorder, and addictions.

Basal Ganglia Basics

The basal ganglia are a collection of interconnected areas deep below the cerebral cortex. They receive information from the frontal cortex about behavior that is being planned for a particular situation. In turn, the basal ganglia affect activity in the frontal cortex through a series of neural projections that ultimately go back up to the same cortical areas from which they received the initial input. This circuit enables the basal ganglia to transform and amplify the pattern of neural firing in the frontal cortex that is associated with adaptive, or appropriate, behaviors, while suppressing those that are less adaptive. The neurotransmitter dopamine plays a critical role in the basal ganglia in determining, as a result of experience, which plans are adaptive and which are not.

Evidence from several lines of research supports this understanding of the role of basal ganglia and dopamine as major players in learning and selecting adaptive behaviors. In rats, the more a behavior is ingrained, the more its neural representations in the basal ganglia are strengthened and honed.[1] Rats depleted of basal ganglia dopamine show profound deficits in acquiring new behaviors that lead to a reward. Experiments pioneered by Wolfram Schultz, M.D., Ph.D., at the University of Cambridge have shown that dopamine neurons fire in bursts when a monkey receives an unexpected juice reward.[2] Conversely, when an expected reward is not delivered, these dopamine cells actually cease firing altogether, that is, their firing rates "dip" below what is normal. These dopamine bursts and dips are thought to drive changes in the strength of synaptic connections—the neural mechanism for learning—in the basal ganglia so that actions are reinforced (in the case of dopamine bursts) or punished (in the case of dopamine dips).

In 1996, Read Montague, Ph.D., and colleagues showed that these patterns of dopamine firing bear a striking resemblance to learning signals developed independently by artificial intelligence researchers.[3] The researchers created a program that would "train" a computer to discover on its own the best sequence of actions needed for the computer program to obtain a simulated reward, such as an endpoint in an artificial maze. The program tries to "predict" when rewards are likely. When this prediction is wrong, the resulting dopaminelike "prediction errors" are used as a learning signal to improve future predictions, which are, in turn, used to modify subsequent actions by the computer. The same computer program has been used for purposes as diverse as learning an optimal strategy for playing computerized backgammon and accounting for foraging behavior in honeybees.

At this point you might be thinking, "Okay, those are rats, monkeys, bees, and automated computer programs. We humans are much more sophisticated than that." But in truth, ample evidence can be found to show that these very same principles of positive and negative reinforcement are relevant in humans—although other cognitive systems complement and can sometimes override this primitive system. Using functional neuroimaging, Samuel McClure, Ph.D., and colleagues at Princeton University showed that humans activated reward areas of the basal ganglia, which are heavily enriched with dopamine, when receiving unexpected rewards. This was true regardless of whether humans received concrete rewards, such as juice (as in the monkey studies), or more-abstract rewards, such as money. Indeed, the same basal ganglia areas were activated even when study participants simply received visual feedback informing them whether they were correct or incorrect in a cognitive task.[4] Further evidence comes from a 1996 study by Barbara Knowlton, Ph.D., and her colleagues at the University of California, Los Angeles. They learned that people with Parkinson's disease, whose basal ganglia dopamine levels are severely depleted as a result of cell death, show specific deficits in exactly this kind of trial-and-error learning from feedback. Thus humans, too, recruit their "primitive" reinforcement learning system in the basal ganglia to support behavior in more-complex cognitive tasks.

How Do the Basal Ganglia Learn?

While these studies and other evidence point to the critical role of the basal ganglia dopamine system in learning about the consequences of one's actions, it is another question altogether to ask "how." Unfortunately, any medical resident or neuroscience graduate student will tell you that the circuitry linking various neural subregions that collectively form the basal ganglia is so complex and seemingly convoluted that trying to piece the puzzle together can make your head spin.

That problem is precisely why the development of computer models is essential. These models enable researchers to simulate various anatomical and physiological pieces of data, using mathematical equations that capture how groups of neurons communicate activity to other neurons within and between brain areas. By incorporating aspects of neuronal physiology and connectivity that are specific to the basal ganglia into a computer model, we can examine what happens when all of this is put together and the computer model is allowed to evolve dynamically as a result of the input it receives.

Many attempts have been made to model basal ganglia function. Although the models have tackled different levels of analysis, from molecular-level to systems-level interactions, several of them have converged on the same core idea: that the architecture of this seemingly convoluted system is particularly well suited to support "action selection"—that is, to implicitly weigh all available options for what to do next and to choose the best one. Intriguingly, the "actions" that can be selected range from simple motor behaviors (for example, touching your pinkies together) to manipulation of information in memory, such as multiplying 42 by 17 in your head. Although these seem like fundamentally different problems, they share the core abstract action selection problem. That is, just as we have to select a single motor action (or sequence of actions) to "beat out" all other possible actions, we also have to know which piece of information is relevant to update and store into memory so that it takes precedence over other, potentially distracting thoughts.

It seems, then, that our ability to think, reason, and manipulate memories evolved from similar mechanisms that allow an animal to perform impressive sequences of motor actions, like when a bird swoops down to catch a fish. The key difference between these cognitive and motor functions may lie in the specializations of the different regions of the frontal cortex and the actions that each encodes. For instance, motor actions are encoded in the motor cortex, whereas a prefrontal cognitive action involves updating information to be actively held in your mind for a period of time, while you continue to process other incoming sensory information. Notably, the circuits that link parts of the basal ganglia to motor cortical areas are structurally identical to those linking other parts of the basal ganglia to the regions of the prefrontal cortex that are used for cognitive processes. Thus the basal ganglia can play a similar role in selecting among both motor and cognitive actions, by interacting with different parts of the frontal cortex.

Finding "Go" and "NoGo"

Building on a large body of earlier theoretical work, my colleagues and I developed a series of computational models that explore the role of the basal ganglia when people select motor and cognitive actions. We have been focusing on how dopamine signals in the basal ganglia, which occur as a result of positive and negative outcomes of decisions (that is, rewards and punishments), drive learning. This learning is made possible by two main types of dopamine receptors, D1 and D2, which are associated with two separate neural pathways through the basal ganglia.[5] When the "Go" pathway is active, it facilitates an action directed by the frontal cortex, such as touching your pinkies together. But when the opposing "NoGo" pathway is more active, the action is suppressed. These Go and NoGo pathways compete with each other when the brain selects among multiple possible actions, so that an adaptive action can be facilitated while at the same time competing actions are suppressed. This functionality can allow you to touch your pinkies together, not perform another potential action (such as scratching an itch on your neck), or to concentrate on a math problem instead of daydreaming.

But how does the Go/NoGo system know which action is most adaptive? One answer, we think (and as you might have guessed), is dopamine. During unexpected rewards, dopamine bursts drive increased activity and changes in synaptic plasticity (learning) in the Go pathway. When a given action is rewarded in a particular environmental context, the associated Go neurons learn to become more active the next time that same context is encountered. This process depends on the D1 dopamine receptor, which is highly concentrated in the Go pathway. Conversely, when desired rewards are not received, the resulting dips in dopamine support increases in synaptic plasticity in the NoGo pathway (a process that depends on dopamine D2 receptors concentrated in that pathway). Consequently, these nonrewarding actions will be more likely to be suppressed in the future.

My colleagues and I demonstrated that a computer model of these Go/NoGo signals (and the spread of these signals through the rest of the basal ganglia circuit) can learn to produce actions that are most likely to lead to reward in the long run.[6] Similarly, the same dopamine reinforcement learning processes can be extended to reinforce cognitive actions that are essential intermediate steps, for example, doing the arithmetic that is needed to achieve the longer-term goal of preparing your taxes. The process also "punishes" distracting thoughts ("What's for dinner?"), allowing you to stay on task. And it enables complex cognitive working memory operations, such as remembering the figures as you multiply them, to be executed more swiftly and efficiently with practice.

Positive and Negative Learners

This theoretical framework, which integrates anatomical, physiological, and psychological data into a single coherent model, can go a long way in explaining changes in learning, memory, and decision making as a function of changes in basal ganglia dopamine. In particular, this model makes a key, previously untested, prediction that greater amounts of dopamine (via D1 receptors) support learning from positive feedback, whereas decreases in dopamine (via D2 receptors) support learning from negative feedback.

To test these ideas, we developed a computer "game" that requires learning from both positive and negative decision outcomes. We tested healthy college students who were given low doses of three different drugs: a drug that enhances the release of dopamine, a drug that reduces the release of dopamine, and a placebo; each student was tested in each of the three conditions. Participants in the study viewed pairs of symbols on a computer screen and were told to choose one of the symbols in the pair by pressing a left or right key on a keyboard. We gave the students no explicit rule for knowing which symbols to select, but after they made their choice they did receive feedback that told them whether the choice was right or wrong, so that they could learn from trial and error.

The trick was that the feedback was somewhat random—it was not always the same for each choice—so it was impossible always to make the right choice. In the most reliable pair, "AB," choosing symbol A led to positive feedback on 80 percent of trials, whereas choosing symbol B led to 80 percent negative feedback. The results of other choices ("CD" and "EF") were less consistent and more random. We hypothesized that all participants would learn to choose A over B, but that they would do so on different bases, depending on which drug they had been given. When dopamine levels are elevated, we hypothesized, participants should learn to choose symbol A, which had received the most positive feedback (that is, they should learn "Go" to A). But they should be relatively impaired in learning to avoid (NoGo) symbol B: the drug-induced elevations in dopamine would prevent dopamine dips that would normally support this negative feedback learning. In contrast, the drug that reduces dopamine levels should lead to reduced Go learning but relatively enhanced NoGo learning for avoiding symbol B.

To distinguish between these choose-A and avoid-B learning strategies, we conducted an additional test phase, in which participants faced the choice of new combinations. Symbol A was re-paired with other, more neutral symbols, and in other trials, symbol B was paired with these same symbols. Participants were told to just use "gut-level" intuition based on their prior learning to make these new choices, and no feedback was provided. We reasoned that to the extent that an individual had learned Go to A, he would reliably choose symbol A in all test pairs in which it was present. Conversely, if he instead learned NoGo to B, he would more reliably avoid symbol B in all test pairs in which it was present.

We found a striking effect of the different dopamine medications on this positive versus negative learning bias, consistent with predictions from our computer model of the learning process. While on placebo, participants performed equally well at choose-A and avoid-B test choices. But when their dopamine levels were increased, they were more successful at choosing the most positive symbol A and less successful at avoiding B. Conversely, lowered dopamine levels were associated with the opposite pattern: worse choose-A performance but more-reliable avoid-B choices. Thus the dopamine medications caused participants to learn more or less from positive versus negative outcomes of their decisions.

These research discoveries raise the intriguing question of whether individual differences in learning from positive versus negative outcomes of decisions can be found even in nonmedicated healthy people. Indeed, although on average our study participants taking the placebo showed roughly equal choose-A and avoid-B performance, individual participants still performed better at one or the other; we refer to these subgroups as positive or negative learners. An initial study showed that these learners differed in the extent to which their brains responded to reinforcement feedback, as measured by brain electrical activity. Negative learners showed greater neural sensitivity to negative feedback. In principle, it is possible that these behavioral learning biases, and their neural correlates, arise from a combination of psychological, cultural, and experiential factors. Still, we reasoned that they could (at least in part) stem from individual differences in basal ganglia dopamine function, which in turn may be controlled by genetics.

Modern genetic techniques enable us to identify individual genes that specifically control basal ganglia dopamine. We therefore collected DNA (using a simple salivary cheek swab) from sixty-nine college students who were tested with the same positive/negative learning procedure.[7] We looked for a gene coding for DARPP-32, a protein known to control basal ganglia dopamine efficacy and previously shown to mediate dopamine D1 receptor effects on synaptic plasticity in animals. Notably, the presence of a common mutation in this gene accounted for a substantial proportion of the relative positive versus negative learning biases in our study participants. Furthermore, a mutation in another gene previously shown to control the density of (NoGo) D2 receptors in the basal ganglia predicted the extent to which participants learned from negative decision outcomes. Together, these results provide more-specific confirmation of our model's suggestion that Go and NoGo learning depends on the D1 and D2 receptors.

How Might This Apply in Parkinson's Disease?

What does this Go and NoGo learning framework suggest for people with basal ganglia dysfunction? Parkinson's disease offers one of the clearest instances of such dysfunction, since the loss of dopamine in the basal ganglia of people with the disease is well understood. Parkinson's therefore provides a good opportunity to test whether the model's account for how these brain systems learn and decide is plausible, and suggests implications for cognitive deficits stemming from the systems' malfunction.

In a seminal paper by Barbara Knowlton (mentioned earlier) and her colleagues, people with Parkinson's were tested with a "weather prediction" task. In this experiment, patients and healthy participants were given four decks of playing cards and were told to guess whether the particular combination of cards in front of them would predict "rain" or "sun." After each guess, feedback let them know whether or not they were correct. The relationship between the cards and the "weather" outcome was complex; no simple rule determined whether any particular card combination would lead to rain or sun, and because the feedback was somewhat random, it was impossible always to predict correctly (much like the actual weather).

Despite being unable to explicitly state the basis of their choices, healthy study participants implicitly integrated the reinforcement feedback over multiple trials and became progressively more accurate at predicting sun or rain. In contrast, people with Parkinson's showed very little evidence of this implicit learning. Subsequent studies demonstrated that this difficulty did not occur if the Parkinson's patients were instead shown the correct answer in every trial and did not have to learn from the consequences of their actions. While it is often assumed that Parkinson's disease affects only motor function, this and several other recent studies confirm that Parkinson's disease is a complex neuropsychiatric condition that clearly has cognitive effects.

To learn to discriminate between subtly different action-outcome contingencies, the basal ganglia appear to require a healthy range of dopamine bursts and dips to support both Go and NoGo learning. Our model suggests that deficits in implicit learning, such as in the weather prediction study, result from a reduced range of dopamine signals. Indeed, when Parkinson's disease was simulated in our computer model (by reducing the level of dopamine in the simulated basal ganglia), the model showed impaired learning in the weather prediction task similar to what was seen in the Parkinson's patients.[8]

One might, then, logically predict that these cognitive learning deficits would be improved by medications that elevate brain dopamine. But while such medication does improve some aspects of cognition, it can actually further impair or even cause some types of learning deficits. This counterintuitive finding is naturally explained by our basal ganglia model. These medications artificially elevate brain dopamine levels and improve motor deficits of the disease, by shifting the balance from too much NoGo to more Go.[9] But this same effect can prevent patients from learning from negative feedback: the dips of dopamine required to learn NoGo are effectively "filled in" by the medication. In essence, the medication prevents the brain from naturally and dynamically regulating its own dopamine levels, which has a detrimental effect on learning, particularly when dopamine levels should be low, as for negative decision outcomes. This notion might explain why some medicated Parkinson's patients develop pathological gambling behaviors, which could result from enhanced learning from gains together with an inability to learn from losses.

To test this idea, we presented people with Parkinson's disease with the same choose-A/avoid-B learning task once while they were on their regular dose of dopamine medication and another time while off it.[10] Consistent with what we predicted, we found that, indeed, patients who were off the medication were relatively impaired at learning to choose the most positive stimulus A, but showed intact or even enhanced learning of avoid-B. Dopamine medication reversed this bias, improving choose-A performance but impairing avoid-B. This discovery supports the idea that medication prevents dopamine dips during negative feedback and impairs learning based on negative feedback.

The Basal Ganglia in Other Disorders

Given this research on Parkinson's disease and the genetic connection with positive and negative learning that we identified, we can be fairly confident that the positive and negative learning biases we have measured are basal ganglia–dependent. This understanding enables us to explore how the same process might play out in people with other disorders that include problems with the basal ganglia, even if those other disorders have more than one neurological underpinning. For example, in attention-deficit/hyperactivity disorder (ADHD), neurobiological research has consistently implicated dopamine deficiency, and stimulant medications such as Ritalin, often used to treat ADHD, act by directly increasing basal ganglia dopamine.

We administered the same computerized learning task that we had used in people with Parkinson's disease to adults diagnosed with ADHD to determine whether dopamine dysfunction could explain their motivational deficits. Again, study participants came into the lab twice, once when they were off medication and once after taking their regular dose

of stimulant medications. In the off-medication state, ADHD participants showed a global reduction in both choose-A and avoid-B performance, compared with healthy control participants.

Of course, we could chalk up this global deficit to other factors,
including a simple lack of attention or motivation to perform well. But the
medication manipulation was more informative: stimulants dramatically
enhanced choose-A performance by an average of 15 percent (making
it identical to that of healthy controls), while having no effect at all on
avoid-B performance, which remained at near-chance levels. We further
showed that the extent to which medications improved positive, relative to negative, feedback learning was correlated with improvements in
other aspects of higher-level cognition, including the ability to pay attention to task-relevant information while ignoring distracting information.

Moreover, these same principles can be applied to understanding
addiction. Researchers now agree that the majority of drugs of abuse act
by hijacking the natural reward system. When an addict snorts cocaine
or smokes a cigarette, he not only experiences a drug-induced high
("reward"), but the associated dopamine bursts act to further stamp
these destructive behaviors into his brain so that they are more likely to
be repeated.

But it is even worse than that. As I mentioned earlier, dopamine
signals are primarily associated with unexpected rewards. Normally, as
people come to expect a reward, an adaptive process prevents dopamine
bursts from occurring when the reward is actually delivered. This process
prevents rewarding behaviors from being "overlearned," possibly so that
they can be overridden or unlearned if their consequences change for
the worse. Unfortunately, drugs of abuse bypass the circuitry that would
normally enable such discounting of expected rewards, and directly
elevate dopamine levels.[11] Consequently, maladaptive drug-taking
behavior is continually strengthened, making it particularly difficult to
override. This enhanced stimulus-response learning also may explain the
high rate of relapse in recovering addicts who encounter cues associated
with taking drugs.

Negative learners may show relatively harmless traits, such as being generally conservative and avoiding risky menu items at a restaurant. But, at the extreme, a focus on errors and negative feedback can lead to obsessiveness and hyperperfectionism. Some evidence from neuroimaging studies suggests that people with obsessive-compulsive disorder (OCD) have a hyperactive error-monitoring system, as revealed by the activity in their brain when they are making mistakes in a cognitive task. But although some evidence for basal ganglia pathology in this disorder has been found, it is not clear why in some cases an overactive error system leads to avoidance of maladaptive behaviors, while in OCD, people tend to repeat the same behaviors again and again. More-elaborate neuro biological, computational, and psychological studies are needed to decipher this and other conundrums associated with maladaptive behaviors in more-complex disorders, including OCD and schizophrenia.

The Power of Implicit Learning

Taken together, this research on the basal ganglia's role in learning raises some provocative—and, to some, perhaps frightening—possibilities. For example, should grade school children be genotyped to determine whether they are more likely to benefit from teaching that emphasizes positive or negative feedback? We know that underachieving students can thrive when placed in a different learning environment. Attention to individual genetic differences, based on the science, could enhance the probability of success by developing a foundation for determining which students are most likely to succeed in which environment. Similarly, if people with ADHD who are receiving a stimulant medication learn well from positive feedback but not negative feedback, that response may imply that they should be motivated to achieve their goals through positive reinforcement. Giving them negative feedback following maladaptive behavior, or threatening punishment for poor behavior, may simply not be a worthwhile strategy.

Many human behaviors can be understood from the perspective of reinforcement learning, even though it does not entirely account for more-complex behaviors. The emerging scientific consensus is that the more-primitive learning system, centered in the basal ganglia, may dictate more of our choices than we usually like to think. The fundamental principles governing action selection and reinforcement in the basal ganglia can also be extended to explain aspects of higher-order decisions and working memory. Indeed, one can think of these more advanced biological circuits as also reflecting a Go/NoGo decision. The choice in this case is between an action directed by the basal ganglia's implicit learning system or one that is a result of more elaborate conscious processing in prefrontal areas of the brain, which can override the more primitive basal ganglia system.

If a common reaction is that what is effectively a series of neuronal reflexes could never do justice to the full glory of human thought and action, then perhaps, as my graduate neuroanatomy professor once said, people just don't have enough respect for reflexes. And nothing can show this more readily than a failure of the system, as in neurological disorders.

CHAPTER 13

Fading Minds and Hanging Chads

Alzheimer's Disease and the Right to Vote

A *Cerebrum* Classic*

by David A. Drachman, M.D.

David A. Drachman, M.D., is a professor and chairman emeritus of the Department of Neurology at the University of Massachusetts Medical Center. His research is on dementia, Alzheimer's disease and related disorders, and the neurology of aging. He can be reached at david.drachman@umassmed.edu.

---

*From Volume 6, Number 1, Winter 2004

As America's population ages, more people will fall victim to illnesses such as Alzheimer's disease that bring progressive and inexorable dementia. What will happen to the millions of men and women who have taken the right to vote for granted, but gradually become cognitively impaired? Inconsistent, often archaic state laws will provide little guidance, says the author; illnesses leading to dementia are seldom recognized in state statutes. Nor can we expect that the votes of the impaired will be randomly distributed, neutralizing their effect on election outcomes, because several significant causes of biased voting may tend to channel their votes in one direction. The author argues that now is the time to fashion realistic guidelines for dealing with the cognitively compromised voter.

Democracy is a form of government that substitutes election by the incompetent many for appointment by the corrupt few.
　　　　　　　　—GEORGE BERNARD SHAW, *MAJOR BARBARA*

MORE THAN ONE OBSERVER during the final days of the 2000 presidential election may have recalled this wry comment by Shaw, as the secretary of state of Florida, the Florida Supreme Court, and finally the Supreme Court of the United States wrangled over how a tiny plurality of votes would determine the outcome of the election for a nation of nearly 290 million people.

There was never a question about the fundamental principle of democracy: that the idealized will of the people must be realized through the alchemy of voting. But did the varied and arcane mechanisms by which voting took place throughout the country accurately transmit that popular will? In the wake of the 2000 election, that issue became the focus of controversy in a legion of editorials and legal briefs.[1] Most agreed that the voting process could and should be improved. The "butterfly ballots" that confused voters into inadvertently designating a third-party candidate should be eliminated; the punch cards that led to dimpled punches or "hanging chads" were deemed unacceptable. Commissions were formed, reports were written; and the voting technology, site of voting, maintenance of voter rolls, poll worker training, computer and telephone access to voter validation information, accessibility for disabled individuals or those with limited English proficiency, and voter education were all scrutinized.

As the mechanics and procedures of voting were probed, however, physicians began to draw attention to another largely unforeseen problem: voting by people with clinically significant dementia. Physicians taking care of patients with Alzheimer's disease and other dementias noticed that many of their patients' caregivers reported that the patient had voted in the 2000 election. Jason Karlawish and his colleagues found that 69 percent of the patients in a dementia clinic at the University of Pennsylvania had voted,[2] compared with 53.7 percent of the voting age population in the commonwealth of Pennsylvania.[3] Brian Ott and his colleagues found that 60 percent of the patients in a Rhode Island clinic for people with dementia had voted[4]—more than the 54.3 percent of the voting age population in that state, and considerably more than the national average of 51.3 percent. In both cases, some of the patients required assistance to vote, some were quite severely demented, and some voted by absentee ballot.

These are studies of small, select populations; one cannot extrapolate, of course, to any reliable estimate of how many people with dementia are voting. But the data do raise the question of whether the number of cognitively impaired potential voters is substantial. Epidemiologic studies in the United States estimate that between 2.5 and 4 million people have Alzheimer's disease or other dementias of varying degrees of severity. Looked at another way, approximately 10 percent of the population of elderly people (those older than age 65), and nearly half of those older than 85, are cognitively impaired—some still quite functional, others not. Having been functioning adults for most of their lives, they would have every expectation of continuing in their civic roles. With a U.S. population of 37 million people older than age 65, and more than 4 million older than 85 (some 4 percent of eligible voters), voting by patients with dementia could have a decisive role in both local and national elections.

How does voting by people with dementia affect our elections and the democratic system? To what extent might exercising—or not exercising—their right to vote distort the electoral process? What are the legal, medical, ethical, and practical aspects of the issue? And what (if anything) needs to be done?

As we shall see, legal precedent in the United States supports the presumption that voting is a fundamental right, but with limits and exceptions. Today, state laws on voting by people with mental incapacitation are riddled with archaic, inconsistent, and largely uninterpretable language, referring to "idiots," "lunatics," and people who are "non compos mentis" as unfit to vote. None of the state constitutions, statutes, or laws refers specifically to people with progressive dementia. This article will consider that large and important group of elderly people. We will see that it is not the diagnosis of Alzheimer's disease[5] or other dementia that should prevent an individual from voting. Instead, it is that person's mental ability or incapacitation, according to a defined standard, that is most relevant to the continuing right to vote.

The Right to Vote—and Its Limits

The overwhelming trend in law and practice in the United States has been to enfranchise one group after another. Over some two centuries, amendments have steadily broadened the right to vote that is enshrined in the U.S. Constitution. Any proposal to limit the franchise of a group runs counter to this strong historical current.

Although today we uphold virtually universal suffrage for adult U.S. citizens, when the Constitution was written the exclusion from voting of women, slaves, and people without property or other qualifications meant that "the people of the several states" entitled to vote were only about 6 percent of the adult male population. Five amendments to the Constitution have extended voting rights to new groups:

• The Fourteenth Amendment guaranteed "equal protection of the laws" to all citizens;

- The Fifteenth Amendment prevented the denial or abridgment of the right to vote "on account of race, color, or previous condition of servitude";

- The Nineteenth Amendment prevented the denial or abridgment of the right to vote on account of sex;

- The Twenty-fourth Amendment prevented abridgment of the right to vote for president, vice president, electors for president or vice president, senators, or representatives on the basis of poll taxes; and

- The Twenty-sixth Amendment lowered the voting age to eighteen.

The Constitution with these amendments might suggest that all adult citizens in the United States had achieved the right to vote and have their votes counted equally. Until 1965, however, states often used so-called literacy tests to discriminate, particularly against black Americans.

In the forty years since then, laws have guaranteed or facilitated actual voting among still other potential voters. The Voting Accessibility for the Elderly and Handicapped Act of 1984 required voting places for federal elections to be physically accessible; the Americans with Disabilities Act of 1990 (ADA) required states to provide assistance for people with physical or mental disabilities; and the "Motor Voter Act" in 1993 made registration to vote widely available through motor vehicle registration offices and all other offices of state-funded programs that provide services to the general public. To facilitate voting under the ADA, people with mental as well as physical disability could be assisted in voting by having instructions explained in simpler language, being accompanied into the voting booth by a friend or family member (other than an employer or union official), and getting assistance from poll workers in casting a ballot.

Who Cannot Vote?

Today, U.S. federal law excludes two groups of people from voting:

children younger than eighteen (almost 80 million) and noncitizens, both legal and illegal or undocumented (almost 30 million). States retain the right to determine whether some individuals not covered under federal law can vote. Two significant groups of individuals are excluded from voting in many states by constitution, law, or statute: felons currently serving sentences (48 states) or former felons (8 states), totaling about 4 million people; and adult citizens with severe mental impairment (44 states).

Unfortunately, the language in state statutes that refer to mental impairment is often archaic. It is certainly neither medically precise nor legally uniform from state to state. Thus, fifteen states do not permit "idiots," "lunatics," or the "insane" to vote. Thirty-two states disallow citizens with "mental incompetence" or "mental incapacity" from voting. Eleven states preclude voting of citizens who are under "guardianship" or "conservatorship" because of mental disabilities. In thirty-seven states and the District of Columbia, anyone who has been found to be incompetent by a court cannot register to vote. In some states, such as New Hampshire, when a person is declared incompetent, legal rights (including voting) are not removed, unless specified by the court. The variation in these laws, and the confusion in trying to implement them, is evident.[6] These differing categories and approaches to mental impairment are scarcely a guide to determining the voting rights of persons with dementia. Yet they are relevant because the validity of excluding individuals with some type of mental disability (although not, to my knowledge, specifically people with dementia) from voting has been tested in the courts.

The New Jersey Constitution, for example, states that "no idiots or insane person shall enjoy the right to suffrage." In 1999, five absentee ballots submitted by residents committed to Trenton Psychiatric Hospital were segregated, unopened, to be counted only if the voter was later deemed competent. Reviewing this action, the appellate division of the superior court determined that by itself residence at a psychiatric hospital was insufficient to sustain a challenge to the right to vote. "Voting is a fundamental right," the court commented, and "the burden of demonstrating that an individual is incompetent requires proof that is clear and convincing."

In Maine, the constitution provides that "persons who are under guardianship for reasons of mental illness" are prohibited from registering to vote or voting in any election. In two recent referendums (1997 and 2000), attempts to remove this language by amending the constitution were defeated by voters. In 2001, three women under guardianship—two with bipolar disorder, one with "intermittent explosive disorder, antisocial personality, and mild organic brain syndrome" (secondary to encephalitis)—challenged this prohibition. All three provided medical or other evidence of understanding the nature and effect of the act of voting and the ability to make an individual choice on the ballot. The U.S. District Court reviewed this action against the State of Maine and determined that this disfranchisement violated the due process and equal protection clauses of the Fourteenth Amendment to the U.S. Constitution.

Although these legal decisions can appear narrow, dealing with particular aspects of state law related to mental impairment and voting, and not directly involving individuals with dementia, several principles emerge rather clearly:

- The courts consistently recognize voting as a fundamental right.

- The states have what is called a "compelling interest" in ensuring that people who vote understand the nature and effect of the act of voting and have the ability to make a choice.

- The burden of demonstrating that an individual is incompetent to vote requires specific proof that is clear and convincing.

- No broad rule about category or status can be used to deprive an individual of the right to vote. Instead, determination of competence to vote must be on an individual basis.

The Special Case of Dementia

How do these principles and precedents apply to questions about the right to vote of people with dementia? The psychiatric and neurologic causes of impaired mental competence are many, ranging from autism to head trauma, multiple sclerosis to bipolar disorder. Each has its specific manifestations, variable course, and severity.

Alzheimer's disease and other progressive dementias (such as fronto-temporal dementia and Lewy body dementia) have certain characteristics that argue for considering these disorders separately when discussing voting rights. For example, although Alzheimer's disease begins gradually, with mild memory and other cognitive problems, eventually and without exception everyone with Alzheimer's who survives sufficiently long will become incapacitated by the disease. Patients with Alzheimer's disease will become increasingly dependent on others, such as caregivers, aides, or institutional helpers.

Under present law, as people with dementia begin to lose memory and other intellectual abilities, certain privileges and rights may be lost. For example, having a driver's license is a privilege that depends—at least theoretically—on the capability to operate a vehicle competently and understand the rules of the road. Only California now requires physicians to report patients with Alzheimer's disease to the Department of Motor Vehicles. A few other states require older drivers to undergo brief tests of competence to renew their licenses. In most states, the observation by police of incompetent or dangerous driving can result in review and the revocation of a license.

Courts can give guardianship or conservatorship to the next of kin (or others) if a person is determined to be incompetent. This guardianship can be limited (for example, to management of financial matters), so that a spouse can manage the finances of a husband or wife incapable of making decisions, while other rights are not transferred. In addition, a durable power of attorney for health care can transfer the right to determine medical management when someone is incapable of making those decisions. Life and death decisions are transferred from the person with Alzheimer's disease or other dementia to that person's next of kin or another responsible person, who is expected to carry out previously stated or written health care decisions, or to use "substituted judgment," interpreting what the intentions of the patient would be if he or she were still competent.

Thus, the right to make financial, property, and life and death decisions—rights absolutely central to the health, wealth, and autonomy of every person—can be legally transferred to another person. By contrast, the right to vote—arguably a right with far less effect on an individual's life—can never be transferred to a surrogate. Even if a person has always voted as a Republican or a Democrat, or always admired or despised a particular candidate, that person's son, daughter, spouse, or guardian cannot cast a proxy vote for what manifestly would have been the person's wishes. Indeed, even if a particular referendum outcome has direct, immediate relevance to the interests of an incompetent person, no one can use substituted judgment to cast the person's vote in favor of that outcome. If no substitute is permitted for the decision and action of the individual whose vote is being cast, then we must decide when an individual with progressive dementia can no longer vote. We must approach this issue with concern both for the rights of the individual and for a potentially significant distortion of the electoral process, with an impact on democracy itself.

When People with Dementia Vote

At first glance, voting by people with dementia might appear to be irrelevant to election outcomes. After all, barely half of eligible adults in the United States exercise their right to vote, even in presidential elections. In 2000, for example, only 51.3 percent of the voting age population, and only 67.5 percent of registered voters, cast votes in the remarkably close presidential election. Among adults who are competent and do vote, many have a limited grasp of the issues or familiarity with the candidates. In short, voting is far from perfect in the United States, and the potential effect of voting by people with dementia could be judged in this light. Moreover, it seems plausible that if people who are mentally incapacitated by advanced dementia did vote, with no understanding of the candidates, issues, or voting process—essentially making a "dartboard" decision—their votes would make little if any difference. Genuinely random voting would end up as statistically meaningless noise in an election. Votes unguided by intentional choice on one side of the ledger would cancel out similarly unguided votes on the other side.

The reality is quite different. There are two significant ways in which the votes of variably incapacitated people with moderate to profound dementia could be shifted in a given direction to alter an election outcome.

Voting Technology

The presidential election in Florida in 2000 underscored the importance of voting technology. Butterfly ballots confused elderly voters, and punch cards with dimpled ballots and hanging chads left election officials unsure of the choice intended. Research has shown that the placement of names on ballots, the design of ballots, and how familiar voters are with a specific voting process can bias voting, including shifting the presumed randomness of uninformed votes.[7] Efforts to eliminate these problems by using touch-screen electronic voting machines instead of lever-operated devices, paper ballots, punch cards, or optical-scan devices introduce new challenges—for both those seeking fair elections and those hoping to influence election outcomes. Votes of incompetent people may no longer be random when voting technology can accidentally, or by manipulation, bias them toward a specific outcome.

Influencing Votes

The most serious concern is that undue influence could be used to produce wholesale bias in the votes of groups of (mostly elderly) cognitively impaired or incapacitated individuals. In 1999, for example, 1.6 million people lived in nursing homes; some 50 to 70 percent of them were cognitively impaired. Still more people with varying levels of dementia are in assisted-living or retirement facilities. Elderly Americans are the population group most likely to vote. According to the U.S. Census Bureau, 79 percent of adults aged sixty-five to seventy-four were registered voters and 72 percent of them voted in the 2000 election. This live population of dependable voters overshadows the numbers of "dead-roll" voters that were used to shift elections in the days of machine politics in our big cities. Cognitively impaired, institutionalized elderly voters could be especially vulnerable to influence, direction, manipulation, or coercion—resulting in a pool of votes that could be used to turn the course of elections.

Nursing homes, for example, typically have no guidelines and no system for evaluating the competence of residents to vote or for determining if they understand the voting process.[8] A substantial population with problems ranging from trouble walking (not a criteria for voting) to advanced Alzheimer's disease (a potentially serious criteria) could be involved in voting activity in such institutional settings. Opportunities for manipulating this large, vulnerable population are considerable. Although the extent of their actual use is unknown, three potential methods stand out.

The first is volunteer assistance with voting. Nursing home residents or elderly people in day care are often bused or taken by van to the polling place. In nursing homes and day care centers, a visit to the polls could be considered another "activity," comparable to bingo, arts and crafts, or a trip to the park. If volunteers from political organizations escort residents and assist in voting, the opportunity for manipulation of the vote is considerable.

In addition, some nursing homes serve as polling places themselves or send residents to neighboring polling places. In these nursing homes, the activities director is typically in charge of guiding residents through the registration and voting process. Except in districts where the board of elections has issued specific guidelines, much of the process is under the control of untrained and unregulated individuals, who have the ability not only to facilitate voting but also to assist the residents directly in voting.

The greatest opportunity for manipulation of voting in nursing homes would seem to be through the use of mail-in or absentee ballots. Incapacitated people could be influenced (or actually have their ballots filled out for them) by someone else, except in the few places where this process is regulated by a board of elections, as in Chicago and certain counties in Maryland. In the absence of such oversight, who knows how, or by whom, these ballots are completed? In Oregon, where all voting is now by absentee ballot, one source estimated that in the 2000 presidential election about 2.5 percent, or 36,000 ballots of the 1.5 million cast, were filled out, signed, or both by someone other than the registered voter.[9]

The Reality Versus the Ideal

U.S. politics has been called a "contact sport." Candidates sometimes seek, obtain, exercise, and retain power in ways that often subvert the democratic ideal. Disallowing of eligible voters, failure or "fixing" of voting machines, stealing of votes, voting "early and often," use of dead rolls, and many other intentional and accidental misadventures affect the outcomes of elections. In 2000, Alaska, for example, had 38,209 more names on its rolls than the number of voting age people; it was estimated that 10 to 20 percent of the names on Indiana voter rolls are bogus; and Arizona, Idaho, Texas, Oklahoma, Utah, and Wisconsin are reported to have up to 20 percent of bogus names. The practice of purchasing votes, particularly on mail-in ballots, in exchange for favors is still reported in parts of the country. Obviously, the risks presented by these and other opportunities of subverting the electoral process cannot be dismissed lightly.

Perhaps in a nation with 10,000 local voting jurisdictions and no federal voting standards, the emergence of a kind of chaotic nonsystem was to be expected. Most people believe, however, that despite its known shortcomings the election process usually works well enough, except when the voting is very close. Our approach as a nation has usually been to make ongoing improvements in the elective process, with adjustments over the decades in the law, including the U.S. Constitution. Can this ameliorative approach meet the challenge of the cognitively impaired American voter?

What Can, or Should, Be Done?

Voting is a fundamental civil right throughout the United States, and the standards for the mental capacity required to exercise that right are low. Nevertheless, almost every state purports to define some threshold for adequate mental capacity, so that people who are truly incapacitated can be excluded from voting. People with progressive dementia, if they live long enough, will inevitably cross the threshold. How can we draw a line that divides those demented persons who still have the mental capacity to vote from those who no longer do? And how do we then translate this definition into specific guidelines for the electoral system— guidelines that will work in the hurly-burly of real elections?

Surely it is not necessary to wait until we know the exact extent of the problem before taking steps to anticipate and prevent abuses. We know that the number of people with dementia is substantial, and there are risks both of depriving them of the right to vote when they still have the cognitive capacity to do so and of manipulating their vote when that capacity has been lost.

The first step is to formulate a clear definition of impaired mental capacity to vote that can be applied to Alzheimer's disease and other progressive dementias. More precise language defining the level of capability necessary to exercise the right to vote is needed in all states. The language used by the U.S. District Court in Maine in 2001 is exemplary: the individual "must understand the nature and effect of the act of voting and be able to make an individual choice on the ballot."

No simple test of this capability is now applied at the time of voting. Guidelines for determining legal "competence" are not directly relevant. The bar must be set lower when determining competence to vote than when determining the ability to manage one's finances. Nursing homes apply no test in connection with voting, and the so-called Minimum Data Set they maintain by law for other purposes is not specifically relevant to ascertaining voting capacity. Polling places have no mechanism for evaluating capacity to vote, nor are election officials trained or competent to make such judgments. At present, election officials, caregivers, or others may provide any level of "assistance" to individuals with mental impairment in registering to vote, voting at a polling place, or completing mail-in ballots. Not even the inability to function at the minimal level necessary to register or cast a vote would necessarily prevent a person with advanced Alzheimer's disease from voting today.

Some jurisdictions have identified procedures for the use of absentee ballots and voting in nursing homes and assisted-living facilities, where (for example, in the city of Chicago and the state of Maryland) specifically trained election judges and poll watchers supervise the voting. Even with these procedures, however, it is not the cognitive capacity to vote that is being ascertained.

To meet this challenge, three steps should be taken to achieve actual effective screening for the minimal cognitive capacity to be able to vote:

1. Develop and test the reliability of a simple screening test for determining capacity to vote. The Maine decision is an appropriate standard: people who understand the nature and effect of the act of voting and are able to make an individual choice on the ballot should have the right to vote. Absent this level of understanding, the capacity to vote should be forfeited.

2. Inform and train election judges and poll watchers in the standard for the capacity to vote and in the use of a simple screening test. The knowledge and expertise to determine who retains the capacity to vote and who has lost it cannot be assumed, and the polling place is the ultimate site where justified decisions must be made.

3. Develop procedures similar to those used in Maryland and Chicago to supervise voting in nursing homes and assisted-living settings. It would be naive to ignore the possibility of unregulated, "wholesale" influencing of votes by partisan helpers among a large incapacitated and dependent population.

As America's population ages, and the number of people with Alzheimer's disease and other dementias increases, we will face some unexpected ethical consequences of our success in prolonging life. One consequence that we can anticipate now is the conflict between guaranteeing the fundamental right to vote in our country and the potential for abuse when a growing number of people with severe mental incapacity are influenced in the exercise of this right. Determining the proper balance is essential.

A free and fair electoral process—the linchpin of a democratic society—often seems troubled. But, as Winston Churchill said, "many forms of government have been tried, and will be tried in this world of sin and woe. No one pretends that democracy is perfect or all-wise. Indeed, it has been said that democracy is the worst form of Government except all those other forms that have been tried from time to time."[10]

CHAPTER 14

The Human Experience of Time

Beyond 9 to 5:
Your Life in Time
by Sarah Norgate
(Columbia University Press 2006; 208 pages, $24.50)

Reviewed by Lynn Nadel, Ph.D.

Lynn Nadel, Ph.D., is Regents Professor of Psychology at the University of Arizona, where his research focuses on memory, spatial cognition, the cognitive effects of stress, the role of sleep in memory, and the nature of mental retardation in Down syndrome. He can be reached at nadel@u.arizona.edu.

TIME, like space, is a central feature of existence. We live in it, use it wisely or not, and ultimately run out of it. For all that, few books for a broad audience have tried to combine what we know about the sociology, psychology, and neurobiology of time. Sarah Norgate attempts to fill this gap in *Beyond 9 to 5: Your Life in Time*, but the book only partially succeeds, perhaps because time itself slips through one's fingers when one tries to get a good grip on it.

Questions About Time

Norgate poses a multitude of questions about time, chapter by chapter, with some being answered rather more usefully than others. For example, the first chapter essentially asks how different cultures use time and provides a broad picture of the quite different ways in which we value and use time around the globe. While I don't understand why the author chose to put this chapter first, it nonetheless provides some fascinating facts. Did you know that people in the United States vastly underestimate how much free time they have and also feel exceedingly pressed for time? Or that people in Europe work the fewest hours, have the most holiday time, and have seen their leisure time increase dramatically in recent years? One might have thought Europeans would feel less pressured by time than Americans do, but, curiously, that is not the case.

How adults use their time throughout the day is the subject of the next chapter, while a separate chapter later on is devoted to babies. As the author points out, all living things organize their activity in some relation to the daily cycle of light and dark. Biological systems abound that reflect this cycle, including, of course, our sleep-wake patterns. Why we sleep, and why we sleep at night, remain hot topics for research, and real-life experiments are under way to test the consequences of changing how we allocate our time between sleeping and waking. More and more people are working and playing at times for which our biology is not prepared, and only (future) time will tell whether this innovation is good or bad. Since it is often inadvisable to act counter to our biology, eating junk food in the wee small hours of the morning might be a bad idea. If that popular activity is a killer, one might expect the proper allocation of time and other healthy habits to lead to an obvious outcome—having more time by living longer. But of course there is more to it than that, since genetics also plays a central role in determining longevity, as the author discusses in the third chapter.

The psychology of time—how we experience it—is the part of the story I would have put first. This, after all, is what motivates our interest in all the rest. Unfortunately, the chapter dealing with this issue is one of the less successful ones in the book, presenting a hodgepodge of observations that do not ever get at the deeper question of how we understand time. How, for example, do we locate an event in the past, and why are we so poor at pinpointing exactly when something happened? What does the way we use language in talking about time tell us about how we conceptualize time? These are but two questions about the general psychology of time that get either short shrift or no attention at all. Instead, the book provides information about anomalies, such as déjà vu and serving life imprisonment. Though interesting, the treatment of these minor issues does not substitute for the discussion of normal time cognition that would have enriched the book.

The brief discussion about infantile amnesia—that time in our youngest lives that we cannot remember—misses an important fact about the brain. While it is true, as the author points out, that considerable postnatal development of the frontal cortex takes place, no mention is made of the well-documented postnatal development of the hippocampus. This brain system, agreed by all neuroscientists to be central to exactly the kinds of memories missing in infantile amnesia, is now thought to become functional only after eighteen to twenty-four months of life, which closely matches the period of absolute infantile amnesia.

This would have been an excellent point in the book at which to say something coherent about time and memory. Norgate does not ignore this topic, but aspects of the relation between time and memory are sprinkled throughout the book, without any synthesis. This omission surprised me, because memory is the cognitive function whose purpose is most explicitly aimed at dealing with the passage of time. Memory allows us to use experiences from the past to behave more effectively in the present. What's more, memory also allows us to project ourselves into the future, to plan for times yet to come. This mental time travel has recently become an exciting area of cognitive neuroscience research. Scientists have learned that the neural systems engaged in mental time travel, in navigating the past and the future, are largely the same as those engaged in navigating the present. A focused discussion of this deep connection between time and memory would have made a useful addition to the book.

Discussion of the physiology and neurology of time and of disorders that affect perception of time comes next, and the two chapters devoted to these topics, though brief, are among the more successful in the book. The connections between time perception and motor function are particularly interesting, as they further the growing body of research demonstrating intimate links between perception and action in a wide variety of neural and cognitive systems. One wishes the author had pursued this line a bit more vigorously. She does explore at some length disorders of timing and how they influence our ability to act effectively in the world. Here, discussions of such conditions as Parkinson's disease, autism, attention deficit disorder, and addiction show how problems within systems responsible for time and timing can have very broad consequences indeed.

An Incomplete Synthesis

Beyond 9 to 5 closes with a chapter devoted to trying to pull the pieces of the earlier narratives together, with mixed success. It is good to point out that we exist in various scales of time simultaneously. We live in the moment, make plans about next week, remember what happened last year, study the history of ancient Rome, and ponder the Big Bang billions of years ago. Pointing out the number and variety of these scales might have been a good place to start. Here again, one learns many fascinating things, such as how the Aymara (an Amerindian tribe living in the Andes) have a reversed sense of time: they think of the past as in front of them and the future as behind. In most other cultures it works the other way—the past is behind and the future lies ahead. But when you consider that the past is knowable and the future is not, the Aymara way seems less strange. This is but one example of how a connection between time, memory, and even language use is stranded without connection to the broader synthesis I hinted at above.

One also learns in this concluding chapter that people in the United States spend more than $11 billion annually feeding their pets, money that could come close to providing basic health and nutrition for most of the world. Yet people in the United States (and other Western democracies), where food is generally available to all, suffer very high rates of obesity and a host of other life- and time-threatening conditions. Norgate sees this disjunct between a culture of plenty, which feeds its pets and feeds itself to the point of widespread disease, and a world in which many people are undernourished as exemplifying how "in affluent societies people's lifestyles are heavily invested in a present time perspective." This is an interesting and important connection, all the more so as we struggle worldwide to find lifestyles that are sustainable both for the planet and for the people and animals that live on it. But this information is buried at the end of a paragraph on page 139.

Much of the last chapter is devoted to the social conditions that lead to inequalities in how long different people get to be alive and in what they are able to do with the time they have. One can only agree with much that Norgate writes here. Certainly the world would be a far better place if children did not have to spend their time as soldiers or slave laborers and if women and men were to spend an equal amount of time on child rearing and housework. How to budget our time better and how to use modern technology to improve the time we have, rather than making it ever more frantic, are also suitable topics for discussion. But in a book that is part of a series called "Maps of the Mind" such issues seem a bit out of place.

While it may not be fair for a reviewer to bemoan the book that didn't get written, as well as review the one that did, I feel it appropriate to tell *Cerebrum* readers that *Beyond 9 to 5* will not likely sate their appetite for knowledge about how the mind and the brain handle time.

CHAPTER 15

Can Our Minds Change Our Brains?

Train Your Mind, Change Your Brain:
How a New Science Reveals Our
Extraordinary Potential to Transform Ourselves
by Sharon Begley
(Ballantine Books 2007; 285 pages, $24.95)

Reviewed by Michael J. Friedlander, Ph.D.

Michael J. Friedlander, Ph.D., is the Wilhelmina Robertson Professor and chairman of the Department of Neuroscience and the director of Neuroscience Initiatives at Baylor College of Medicine in Houston, Texas. His research focuses on synaptic plasticity and development of the visual cortex. He can be reached at friedlan@bcm.edu.

IN OCTOBER 2004, *Wall Street Journal* science editor Sharon Begley attended a meeting in an improbable location on a seemingly equally improbable topic. At the Dalai Lama's private compound in Dharamsala, India, leading neuroscientists and Buddhist philosophers met to consider "neuroplasticity." The conference was organized by the Mind and Life Institute as part of a series of meetings, beginning in 1987, for brain researchers and Buddhist scholars to share insights into the workings of the mind and the brain. The 2004 meeting set out to answer two questions: "Does the brain have the ability to change, and what is the power of the mind to change it?"

In *Train Your Mind, Change Your Brain: How a New Science Reveals Our Extraordinary Potential to Transform Ourselves*, Begley reveals the results of that unlikely meeting, while making accessible the rapidly emerging science of neuroplasticity. Although the title might suggest otherwise, Begley's book is not a manual on brain exercises or the power of positive thought. Instead, it is a lively, largely scientifically accurate, and eminently readable view into the brain's capacity for malleability. Moreover, the book makes it obvious why the two seemingly disparate cultures of neuroscience and Buddhism share a mutual interest, as well as have much to learn from each other.

Neuroplasticity encompasses all of the things that nerve cells can do (or that can be done to them) to change the structure and/or function of the organ in which they are embedded, the brain. Examples include the brain's ability to add new nerve cells (neurogenesis), change the efficiency by which one neuron talks to another at sites of chemical communication (synaptic plasticity), and remap functional connections to allow for the outputs (axons) of one set of neurons to invade territory vacated by another set of axons. All of these processes, which take place in full force while the brain is initially assembled in utero and during the early years of life, are now known also to occur, to some degree, throughout the life span. This latter type of neuroplasticity is at the heart of the meeting held in Dharamsala and of Begley's book—not just the nuts and bolts of how brains effect this plasticity, but its implications for who we are today (versus yesterday) and our potential to become something else tomorrow.

Discovering Neuroplasticity

The interface of Western science and Buddhist philosophy provides the context for considering these implications. In order to build a bridge between the two disciplines, Begley takes readers on a journey through how neuroscience's view of the brain's capacity for functional reorganization has evolved. She highlights the theories and discoveries of many of the field's leading investigators. Included are the famous late-nineteenth-century psychologist William James; the father of cellular anatomy of the nervous system, Santiago Ramón y Cajal; and modern experimental neuroscientists such as Nobel laureates David H. Hubel, M.D., and Torsten N. Wiesel, M.D., who elegantly demonstrated the sensitivity of cerebral cortical neurons to visual experience during "critical periods" of early brain development. Begley implicitly makes the point that the emphasis on such limited windows of opportunity for functional brain reorganization may have contributed to researchers' initial reluctance to embrace the growing body of evidence for neuroplasticity in the mature brain.

But then she goes on to describe a growing list of research that demonstrates this neuroplasticity, such as the discovery by Michael Merzenich, Ph.D., and Jon Kaas, Ph.D., that the part of the cerebral cortex that processes tactile information is altered after training or when sensory input from a part of the body is removed. The case for lifelong neuroplasticity becomes even more compelling as Begley reports the identification by Fred H. Gage, Ph.D., of neurogenesis in the brains of adult mammals. The reader will be intrigued to learn that many of the insights into the capacity for adult neuroplasticity came from unexpected quarters. For example, lifelong neuronal renewal was first established in songbirds, and the initial observations of human adult neurogenesis came from cancer patients who underwent chemotherapy with drugs that serendipitously allowed visualization of newly generated neurons when their brains were examined by microscope upon autopsy.

These and other discoveries complete the plasticity circle: not only can the brain change the functional efficiency of synaptic connections, sprout new connections, and refine existing connections, but it also can essentially substitute or replace at least some of its component parts, not just in the early stages of development but in adulthood as well. We learn through Begley's narrative how the sands of the collective contemporary scientific wisdom shift, grudgingly at first but then decisively under the weight of increasing evidence.

Can Our Minds Change Our Brains?

All the while, the case for adult neuroplasticity leads us back to the larger questions raised in Dharamsala, ones that have challenged philosophers for millennia. A long-standing question to which Begley introduces readers is the concept that the mind can be viewed as a result of the processes that brains carry out. But new questions are brought by modern neuroscience: Are we the same person after we reconfigure our brain hardware? What does the stuff of brain mean for our mind? If the outer world can change our brains (through sensory experience, injury, exercise), can we change our brains through internal commands—that is, can we will our mind to change the very matter on which the mind runs? Even though parts may come and go, the processes and operations performed by the parts have a history (memory), with lifelong plasticity being part of who we are.

This plasticity sheds new light on the interplay between nature and nurture. Begley weaves the harrowing story of the children in Romanian orphanages under the Ceausescu regime together with studies by Michael Meaney, Ph.D., at McGill University on how maternal attention affects gene expression and behavior in rat pups:

By the time the children were six, a 2004 study of Romanian children adopted by British families concluded, there were major persistent deficits in a substantial minority of them. The scientists attributed that to some form of early biological programming or neural damage stemming from institutional deprivation. It may seem a long way from Romania's abandoned children to the rats in cages at McGill University in Montreal. But neuroscientist Michael Meaney begs to differ. ... The brains of rats born to neglectful mothers but raised by high licking attentive ones had as many glucocorticoid receptors as rats born to and raised by high licking mothers. ... The mother rats literally groom their offspring to have the adult temperaments and mothering styles they do. ... Score another point for nurture over nature. The genes of the mellow rats are identical to the genes of the neurotic rats, at least in terms of how genetics traditionally defines "identical"—the sequence of molecules that were the holy grail of the Human Genome Project. But this sequence does not represent nature's orders. It's more like a suggestion. Depending on what sort of world a creature finds itself in, that sequence might be silenced or amplified, its music played or muted, with diametrically different effects on behavior and temperament.

So how is all of this fascinating information about the interplay between behavior, genes, and neuroplasticity connected to the questions posed by the Buddhist philosophers about who we are and how our minds might affect our brains? At the November 2005 annual meeting of the Society for Neuroscience, thousands of attendees at a lecture on the neuroscience of meditation asked that same question as they listened to the Dalai Lama discuss how Buddhism can interface with the study of the human condition. The connection becomes clearer as the reader of Begley's book learns about the Dalai Lama's lifelong personal fascination with science, Buddhism's goal of insight, and its rich consonances with neuroscience. Begley explores how Buddhism may both learn from and inform the search for answers to the great questions about the human condition.

Scientists are now clear that sensory input from the outside world can change the organization of synaptic networks within the brain (often called bottom-up plasticity). Interestingly, when active attention or motor behavior is coupled with that sensory input, functional reorganization, including synaptic plasticity, is enhanced. But what about processes generated entirely within our brains, without any physical interaction with the environment, such as sensory input or motor output? Can such processes alone physically change the very brains in which they are operating (top-down plasticity, or "mind over matter")? This question is of great importance to neuroscientists, to psychiatrists and their patients, and to followers and scholars of Buddhism.

Is simply imagining the desired outcome, for example mental rehearsal by an athlete or musician, sufficient to induce plasticity? Might this ability be used to help fix a malfunction in the brain of someone with a psychiatric disorder such as depression? In *Train Your Mind, Change Your Brain*, Begley makes the case for bidirectional interaction. This raises the intriguing issue of whether internal training—including Buddhist meditation—might trigger the biological processes of plasticity, perhaps making wholesale changes in neuronal architecture, triggering neurogenesis, or healing our own injured or diseased brain.

This part of the book motivates the reader to confront big issues, including the uniformity versus duality of the mind/brain. Begley observes that some scientists argue for the mind being more than the brain's physical activity, but that the corollary of this argument—that the mind can change the brain—is particularly interesting. However, she goes on to write: "A mental state, be it a sense of the color red or the sound of B-sharp or the emotion of sadness or the feel of pain, is more than its neural correlates." But even though we are still woefully uninformed as to the nature of those processes of mentation, including perception and consciousness, we have plenty of opportunity for biological exploration of these processes' physical nature. This exploration eventually will go beyond the capability of current tools, such as analysis of patterns of nerve impulses and blood oxygen signals, in order to discover the physical processes that may very well be the computational operations of mind.

Overall, Begley provides a clear light along the scientific trail from immutable brains to lifelong neuroplasticity. This history is presented accurately and fairly, while being melded with the big questions of mind as engaged by a major philosophy, Buddhism. The otherwise informative and well-framed thesis is somewhat weakened, however, by a perhaps overly optimistic picture of the expression of neuroplasticity throughout life. The reader who is not aware of the current controversies in this field may come away with the mistaken impression that plasticity in the adult nervous system is equal to that early in life and perhaps not mechanistically constrained.

For example, after a raging controversy over whether the capacity for adult neurogenesis is ubiquitous, the process is currently understood as manifesting only in select brain regions (particularly the dentate gyrus and olfactory bulb, but not in the neocortex). In a similar vein, training-induced changes in the sensory and motor map organization of the cerebral cortex are more robust in young brains. Moreover, the capacity for fairly rapid functional reorganization of the adult visual cortex after retinal injury has recently been challenged. In some cases, Begley points out differences between young and mature brains, but the jury is still out on the degree of adult plasticity that can occur in some of these systems.

In general, though, the author's points regarding lifelong neuroplasticity are accurate, cleverly described for a broad base of readers, and woven into an at once entertaining and informative book. Begley takes us on a journey that will open readers' eyes to the enormous adaptive capacity of our brains. *Train Your Mind, Change Your Brain* provides a firm foundation for a continuing and expanding dialogue between people who usually look at the world through very different lenses.

CHAPTER 16

Seeking Insights into the Human Mind in Art and Science

Proust Was a Neuroscientist
by Jonah Lehrer
(Houghton Mifflin Company 2007; 230 pages, $20.00)

Reviewed by Steven Rose, Ph.D.

Steven Rose, Ph.D., is emeritus professor of biology at Britain's Open University, of which he was one of the founders in 1969, and a visiting professor at University College London. His research, on which he has published some 300 papers and reviews, focuses on the molecular and cellular mechanisms of memory and currently on a potential treatment for memory loss in Alzheimer's disease. He is the author of many popular books about science, including *The Making of Memory, Lifelines: Biology Beyond the Gene,* and *The Future of the Brain.*

PROUST WAS A NEUROSCIENTIST? No, despite Jonah Lehrer's provocative title, the novelist Marcel Proust was not.

Proust's seven-volume novel, *À la recherche du temps perdu* (English translations are titled either *Remembrance of Things Past* or *In Search of Lost Time*), published between 1913 and 1927, is a profound meditation on the nature of emotional and sensual memory and the complex interpersonal relationships of a decadent aristocracy and a rising bourgeoisie. Researchers studying memory will almost certainly be aware of the famous passage, early on in the first volume, where the taste of a madeleine cake evokes in Proust's semiautobiographical narrator an entire ensemble of childhood memories, as it is one of the few references to the work of a novelist to find its way regularly into neuroscience textbooks. But while Proust was profoundly introspective and focused on his own thoughts and feelings, his concern with the bodily mechanisms that underlay them was almost certainly confined to medical consultations about his perennially poor health.

Lehrer's title thus reflects both the ambitious goals of his book and their limitations. His thesis, presented in a series of eight case studies, is that through the nineteenth and early twentieth centuries, writers, painters, musicians, and even cooks achieved insights into the mind that both contradicted the assumptions of the sciences of their time and anticipated some of the understanding of the brain that modern neuroscience offers. It's a fun and thought-provoking argument, even though I feel that at times his case remains at best nonproven.

Did Artists Anticipate Modern Brain Science?

It has become increasingly fashionable to explore the links between art and science. I myself have had the privilege of being "shadowed" in the laboratory and at conferences for a year by a novelist who used the experience in her story of a neuroscientist who was researching Alzheimer's disease as he confronted his own past as a Holocaust survivor. In the main, such projects all have a similar intent: to expose artists, novelists, and musicians to modern sciences ranging from cosmology to molecular biology, and by so doing to stimulate their creative juices. The process may be enjoyable for the scientists involved, but it is not normally supposed to affect their research programs.

In arguing that art anticipates science, Lehrer—a writer and editor who has served his time as a technician in Eric Kandel's memory research laboratory and in the kitchens of distinguished restaurants—takes the opposite approach. *Proust Was a Neuroscientist* focuses on Walt Whitman, George Eliot, Auguste Escoffier, Marcel Proust, Paul Cézanne, Igor Stravinsky, Gertrude Stein, and Virginia Woolf. Each artist is supposed to exemplify a change in how we think about mind and emotion. In Lehrer's account, these insights are the biology of, respectively, feeling, freedom, taste, memory, sight, sound, language, and the self. Lehrer's choice of subjects is eclectic and fascinating, and the brief biographies and summaries of their work that he provides are enticing samplers for newcomers. The work of each artist is given a counterweight in terms of a relevant aspect of modern neuroscience. Thus his account of Proust is followed by a review of research in the Kandel lab (in some of which Lehrer participated) on the molecular mechanisms involved in memory formation and the ways in which memories can be reactivated by appropriate stimuli.

The challenges of Lehrer's exercise should be immediately apparent. His accounts of the intellectual breakthroughs made by his chosen authors and artists risk trespassing on the oft-plowed fields of generations of literary critics and historians of science. At the same time, they may raise the ire of neuroscientists who feel that he has simplified their findings to suit his thesis. But a brash twenty-five-year-old such as Lehrer often dares to tread where older and more cautious people might not, and in doing so he may provide left-field insights that others have missed. So even if Lehrer's art-science linkages sometimes seem stretched to the breaking point, the attempt to make them is itself thought-provoking.

Examining the Examples

I couldn't help but quarrel both with some of Lehrer's choices and with his interpretations of their work. For example, the chef Escoffier is said by Lehrer to have invented making stock by boiling down bones, anticipating the Japanese discovery of umami, the so-called fifth taste. Lehrer recounts Escoffier's story with such gusto as to whet the reader's appetite in a more than usually literal sense (his time in restaurant kitchens was clearly not wasted). Yet my facsimile copy of the classic cookbook *Mrs. Beeton's Book of Household Management*, published in 1861, a generation before Escoffier, has just such a recipe and reflection on the culinary virtues of stock, and certainly the Japanese knew about stock long before its tasty effect was linked to the presence of the amino acid glutamate. But maybe a dispute about precedence and historical accuracy doesn't matter to the general argument about the fascinating relationships among cooking, biochemistry, and the experience of taste.

Similarly, the choices of Cézanne for sight and Stravinsky for sound make sense, though equal cases could have been made for Pablo Picasso and Arnold Schoenberg. Individual artists, like individual scientists, write, paint, and research in a social context that helps shape both the problems they set out to resolve and the ways they approach them, so it is not surprising to find similarities among contemporaries.

I have more serious problems over George Eliot. She is an excellent choice as subject, since she is the only one of Lehrer's exemplars who was seriously interested in science in general and biology in particular. To the scandal of Victorian society, she lived openly with the already married biologist George Lewes. She was fascinated by the implications of Darwinian evolutionary theory (it has been argued that at least one of her novels, *Mill on the Floss*, is structured as a meditation on evolution). In her greatest novel, *Middlemarch*, a young country doctor tries to develop a medicine based on current science and innovates by using a stethoscope and microscope. But to locate Eliot as the discoverer of "the biology of freedom" is to ignore that almost all her novels, including *Middlemarch* (from which Lehrer quotes extensively), are concerned with the lack of freedom for women. He even regards that novel as having a happy ending, ignoring the painful irony of its closing words, in which the frustrated heroine, Dorothea, settles for the best that can be expected for a woman: marriage to a nondescript but handsome artist.

As to the neuroscience of freedom that Eliot is supposed to have anticipated, it is almost comically inadequate to discover that Lehrer locates it in brain plasticity and in the recognition that the long-held belief that we emerge into the world with a full complement of neurons, which steadily die off as we age, is wrong. While it is now clear that even in older adults, neurons are steadily being born in key brain regions, the implications of this important and exciting research finding are still uncertain. Yet I suspect that Eliot would have been as surprised by being linked to adult neurogenesis as Gertrude Stein would be to find herself, as indeed she is, matched by Lehrer with the linguist Noam Chomsky.

Although Lehrer inexplicably misses her passionate feminism, Virginia Woolf is an excellent choice for understanding the self, as her writings, like those of James Joyce, explore her subjects' interior thoughts rather than their external actions. Indeed, this was a recurrent theme at the beginning of the twentieth century, as issues of the self and identity became important—think too of Freud. The writer and critic David Lodge's 2002 book *Consciousness and the Novel* explores the way in which, from the purely exterior narrative of eighteenth-century novels through the great nineteenth-century writers, authors entered deeper and deeper into their creations' interior lives. Woolf's subjects, Mrs. Ramsay and Mrs. Dalloway, exist almost entirely in such an interior world. But I would have paired her with Antonio Damasio, as the neuroscientist who has written most penetratingly about the emergence of different levels of self, rather than Roger Sperry, with his studies of dual consciousness in patients whose corpus callosum was severed, and Christof Koch's search for the neural correlates of consciousness. It is in response to the current enthusiasm among neuroscientists for such brain theories of consciousness that Lodge argues, in a way that would surely appeal to Lehrer, that novelists can still reach deeper into the experience of consciousness than any of us practicing at the laboratory bench.

Going Beyond Scientific Reductionism

This argument, in fact, is why Proust wasn't a neuroscientist. With the exception of Eliot, the artists and writers Lehrer chooses, although fascinated by mind and the senses, were pretty indifferent to the brain processes that might underlie them, whereas it is precisely these brain processes that are the concern of us neuroscientists, however much we may wish to extrapolate our findings to "explain" the mind.

Over and over again throughout the book, Lehrer recognizes that understanding the mind requires going beyond reductionism, however powerful a research tool that way of thinking remains. By encouraging neuroscientists to explore the worlds and thoughts of his chosen subjects, Lehrer may thus help to enrich our own research programs, just as the various more conventional art/science projects are supposed to enrich the work of the artists who participate in them. All of us need to understand how it is that the brain is embodied, and how we, as owners of both brain and body, are embedded in the surrounding social world. Writers, artists, and musicians can still tell us things about ourselves that all our genetic wizardry, molecular tricks, probing electrodes, and magnetic brain imaging cannot approach.

Endnotes

2. REMEMBERING THE PAST TO IMAGINE THE FUTURE

1. Endel Tulving, "Memory and Consciousness," *Canadian Psychologist* 26 (1985): 1–12.
2. Stanley B. Klein, Judith Loftus, and John F. Kihlstrom, "Memory and Temporal Experience: The Effects of Episodic Memory Loss on an Amnesic Patient's Ability to Remember the Past and Imagine the Future," *Social Cognition* 20 (2002): 353–79.
3. Demis Hassabis, Dharshan Kumaran, Seralynne D. Vann, and Eleanor A. Maguire, "Patients with Hippocampal Amnesia Cannot Imagine New Experiences," *Proceedings of the National Academy of Sciences USA* 104 (2007): 1726–31.
4. Janie Busby and Thomas Suddendorf, "Recalling Yesterday and Predicting Tomorrow," *Cognitive Development* 20 (2005): 362–72.
5. Arnaud D'Argembeau and Martial Van der Linden, "Phenomenal Characteristics Associated with Projecting Oneself Back into the Past and Forward into the Future: Influence of Valence and Temporal Distance," *Consciousness and Cognition* 13 (2004): 844–58.
6. Karl K. Szpunar, Jason M. Watson, and Kathleen B. McDermott, "Neural Substrates of Envisioning the Future," *Proceedings of the National Academy of Sciences USA* 104 (2007): 642–47.
7. Donna Rose Addis, Alana T. Wong, and Daniel L. Schacter, "Remembering the Past and Imagining the Future: Common and Distinct Neural Substrates During Event Construction and Elaboration," *Neuropsychologia* 45 (2007): 1363–77.
8. Hermann Ebbinghaus, Memory: A Contribution to Experimental Psychology, trans. Henry A. Ruger and Clara E. Bussenius (New York: Dover, 1964). Originally published as *Über das Gedächtnis: Untersuchungen zur experimentellen Psychologie* (Leipzig: Duncker & Humblot, 1885).

4. PROTECTING THE BRAIN FROM A GLUTAMATE STORM

1. Chrysanthy Ikonomidou and Lechoslaw Turski, "Why Did NMDA Receptor Antagonists Fail Clinical Trials for Stroke and Traumatic Brain Injury?," *Lancet Neurology* 1, no. 6 (2002): 383–86.
2. Miroslav Gottlieb, Yin Wang, and Vivian I. Teichberg, "Blood-Mediated Scavenging of Cerebrospinal Fluid Glutamate," *Journal of Neurochemistry* 87 (2003): 119–26.
3. Alexander Zlotnik, Boris Gurevich, Sergei Tkachov, Ilana Maoz, Yoram Shapira, and Vivian I. Teichberg, "Brain Neuroprotection by Scavenging Blood Glutamate," *Experimental Neurology* 203 (2007): 213–20.

6. RISKS AND REWARDS OF BIOLOGICS FOR THE BRAIN

1. Chris H. Polman, Paul W. O'Connor, Eva Havrdova, et al., "A Randomized, Placebo-Controlled Trial of Natalizumab for Relapsing Multiple Sclerosis," *New England Journal of Medicine* 354 (2006): 899–910.
2. Stephen L. Hauser and S. Claiborne Johnston, "Recombinant Therapeutics: From Bench to Bedside (If Your Health Plan Concurs)," *Annals of Neurology* 60 (2006): 10A–11A.
3. Gary Walsh, "Biopharmaceutical Benchmarks 2006," *Nature Biotechnology* 24 (2006): 769–76.
4. Ibid.
5. Ibid.

7. "COSMETIC NEUROLOGY" AND THE PROBLEM OF PAIN

1. Anjan Chatterjee, "Cosmetic Neurology: The Controversy over Enhancing Movement, Mentation, and Mood," *Neurology* 63 (2004): 968–74.
2. President's Council on Bioethics, *Beyond Therapy: Biotechnology and the Pursuit of Happiness* (New York: ReganBooks, 2003).
3. Joseph E. LeDoux, "Emotion Circuits in the Brain," *Annual Review of Neuroscience* 23 (2000): 155–84.
4. James L. McGaugh, "The Amygdala Modulates the Consolidation of Memories of Emotionally Arousing Experiences," *Annual Review of Neuroscience* 27 (2004): 1–28.
5. Donald Caton, *What a Blessing She Had Chloroform: The Medical and Social Response to the Pain of Childbirth from 1800 to the Present* (New Haven, Conn.: Yale University Press, 1999).
6. Ibid.
7. Ibid.
8. Donald D. Price, "Psychological and Neural Mechanisms of the Affective Dimension of Pain," *Science* 288 (2000): 1769–72.
9. Ronald C. Kessler, Wai Tat Chiu, Olga Demler, and Ellen E. Waters, "Prevalence, Severity, and Comorbidity of 12-Month *DSM-IV* Disorders in the National Comorbidity Survey Replication," *Archives of General Psychiatry* 62 (2005): 617–27.

8. WHEN MUSIC STOPS MAKING SENSE: LESSONS FROM AN INJURED BRAIN

1. Ian McDonald, "Musical Alexia with Recovery: A Personal Account," *Brain* 129 (2006): 2554–61.
2. Daniele Schön, Carlo Semenza, and Gianfranco Denes, "Naming of Musical Notes: A Selective Deficit in One Musical Clef," *Cortex* 37, no. 3 (2001): 407–21.
3. A. Bevan, G. Robinson, B. Butterworth, and L. Cipolotti, "To Play 'B' but Not to Say 'B': Selective Loss of Letter Names," *Neurocase* 9, no. 2 (2003): 118–28.

4. Daniele Schön, Jean Luc Anton, Muriel Roth, and Mireille Besson, "An fMRI Study of Music Sight-Reading," *NeuroReport* 13, no. 17 (2002): 2285–89; Justine Sergent, Erik Zuck, Sean Terriah, and Brennan MacDonald, "Distributed Neural Network Underlying Musical Sight-Reading and Keyboard Performance," *Science* 257, no. 5066 (1992): 106–9.

5. Marcus E. Raichle and Debra A. Gusnard, "Intrinsic Brain Activity Sets the Stage for Expression of Motivated Behavior," *Journal of Comparative Neurology* 493, no. 1 (2005): 167–76.

6. Isabelle Peretz and Max Coltheart, "Modularity of Music Processing," *Nature Neuroscience* 6, no. 7 (2003): 688–91.

9. SEEKING FREE WILL IN OUR BRAINS

1. Mark Hallett, "Volitional Control of Movement: The Physiology of Free Will," *Clinical Neurophysiology* 118 (2007): 1179–92.

2. Benjamin Libet, Curtis A. Gleason, Elwood W. Wright, and Dennis K. Pearl, "Time of Conscious Intention to Act in Relation to Onset of Cerebral Activity (Readiness-Potential): The Unconscious Initiation of a Freely Voluntary Act," *Brain* 106 (1983): 623–42.

3. See note 1 above.

4. Hakwan C. Lau, Robert D. Rogers, and Richard E. Passingham, "Manipulating the Experienced Onset of Intention After Action Execution," *Journal of Cognitive Neuroscience* 19 (2007): 81–90.

5. Marlene R. Cohen and William T. Newsome, "What Electrical Microstimulation Has Revealed About the Neural Basis of Cognition," *Current Opinion in Neurobiology* 14 (2004): 169–77.

10. STRESS AND IMMUNITY: FROM STARVING CAVEMEN TO STRESSED-OUT SCIENTISTS

1. Herbert Herzog, "Neuropeptide Y and Energy Homeostasis: Insights from Y Receptor Knockout Models," *European Journal of Pharmacology* 480, nos. 1–3 (2003): 21–29.

2. Michael W. Schwartz and Randy J. Seeley, "Seminars in Medicine of the Beth Israel Deaconess Medical Center: Neuroendocrine Responses to Starvation and Weight Loss," *New England Journal of Medicine* 336, no. 25 (1997): 1802–11.

3. Paul A. Baldock, Amanda Sainsbury, Michelle Couzens, et al., "Hypothalamic Y2 Receptors Regulate Bone Formation," *Journal of Clinical Investigation* 109, no. 7 (2002): 915–21.

4. Julie Wheway, Charles R. Mackay, Rebecca A. Newton, et al., "A Fundamental Bimodal Role for Neuropeptide Y1 Receptor in the Immune System," *Journal of Experimental Medicine* 202, no. 11 (2005): 1527–38.

12. "GO" AND "NOGO": LEARNING AND THE BASAL GANGLIA

1. Mandar S. Jog, Yasuo Kubota, Christopher I. Connolly, Viveka Hillegaart, and Ann M. Graybiel, "Building Neural Representations of Habits," *Science* 286 (1999): 1745–49.
2. Wolfram Schultz, "Getting Formal with Dopamine and Reward," *Neuron* 36 (2002): 241–63.
3. P. Read Montague, Peter Dayan, and Terrence J. Sejnowski, "A Framework for Mesencephalic Dopamine Systems Based on Predictive Hebbian Learning," *Journal of Neuroscience* 16 (1996): 1936–47.
4. Carol A. Seger and Corinna M. Cincotta, "The Roles of the Caudate Nucleus in Human Classification Learning," *Journal of Neuroscience* 25 (2005): 2941–51.
5. Jonathan W. Mink, "The Basal Ganglia: Focused Selection and Inhibition of Competing Motor Programs," *Progress in Neurobiology* 50 (1996): 381–425.
6. Michael J. Frank, "Dynamic Dopamine Modulation in the Basal Ganglia: A Neurocomputational Account of Cognitive Deficits in Medicated and Nonmedicated Parkinsonism," *Journal of Cognitive Neuroscience* 17 (2005): 51–72.
7. Michael J. Frank, Ahmed A. Moustafa, Heather M. Haughey, Tim Curran, and Kent E. Hutchison, "Genetic Triple Dissociation Reveals Multiple Roles for Dopamine in Reinforcement Learning," *Proceedings of the National Academy of Sciences* 104 (2007): 16311–16.
8. See note 6 above.
9. See note 5 above.
10. Michael J. Frank, Lauren C. Seeberger, and Randall C. O'Reilly, "By Carrot or by Stick: Cognitive Reinforcement Learning in Parkinsonism," *Science* 306 (2004): 1940–43.
11. A. David Redish, "Addiction as a Computational Process Gone Awry," *Science* 306 (2004): 1944–47.

13. FADING MINDS AND HANGING CHADS: ALZHEIMER'S DISEASE AND THE RIGHT TO VOTE

1. U.S. Commission on Civil Rights, *Voting Rights in Florida 2002: Briefing Summary*, August 2002, www.usccr.gov/pubs/vote2000/sum0802.htm (accessed January 2004).
2. Jason H. T. Karlawish, David A. Casarett, Bryan D. James, Kathleen J. Propert, and David A. Asch, "Do Persons with Dementia Vote?," *Neurology* 58 (2002): 1100–1102.
3. Federal Election Commission, "Voter Turnout in 2000," www.fec.gov/pages/2000turnout (accessed January 2004).
4. Brian R. Ott, William C. Heindel, and George D. Papandonatos, "A Survey of Voter Participation by Cognitively Impaired Elderly Patients," *Neurology* 60 (2003): 1546–48.

5. Guy McKhann, David Drachman, Marshall Folstein, Robert Katzman, Donald Price, and Emanuel M. Stadlan, "Clinical Diagnosis of Alzheimer's Disease: Report of the NINCDS-ADRDA Work Group Under the Auspices of Department of Health and Human Services Task Force on Alzheimer's Disease," *Neurology* 34 (1984): 939–44.

6. Kay Schriner and Lisa Ochs, "'No Right Is More Precious': Voting Rights and People with Intellectual and Developmental Disabilities," *Policy Research Brief* (University of Minnesota, Minneapolis, Institute on Community Integration) 11, no. 1 (2000).

7. R. Michael Alvarez, "Ballot Design Options" (paper prepared for the Center for American Politics and Citizenship project "Human Factors Research on Voting Machines and Ballot Designs: An Exploratory Study," February 17, 2002), www.capc.umd.edu/repts/MD_EVote_Alvarez.pdf (accessed December 2003).

8. Joan L. O'Sullivan, "Voting and Nursing Home Residents: A Survey of Practices and Policies," *Journal of Health Care Law and Policy* 4, no. 2 (2001): 325–53.

9. *Los Angeles Times*, December 11, 2000.

10. Winston Churchill, speech delivered in the House of Commons, London, November 11, 1947.

Index

Other Dana Press Books

www.dana.org/books/press

Books for General Readers

Brain and Mind

YOUR BRAIN ON CUBS: Inside the Heads of Players and Fans

Dan Gordon, Editor

Neuroscientists and science writers explore how the brain relates to questions such as: What makes fans loyal? What is involved in seeing the ball, deciding ball or strike, making a decision to swing? What allows us to believe in a "curse"?

6 illustrations.

Cloth • 150 pp • ISBN-13: 978-1-932594-28-7 • $19.95

THE NEUROSCIENCE OF FAIR PLAY:
Why We (Usually) Follow the Golden Rule

Donald W. Pfaff, Ph.D.

A distinguished neuroscientist presents a rock-solid hypothesis of why humans across time and geography have such similar notions of good and bad, right and wrong.

10 illustrations.

Cloth • 234 pp • ISBN-13: 978-1-932594-27-0 • $20.95

BEST OF THE BRAIN FROM SCIENTIFIC AMERICAN:
Mind, Matter, and Tomorrow's Brain

Floyd E. Bloom, M.D., Editor

Top neuroscientist Floyd E. Bloom has selected the most fascinating brain-related articles from *Scientific American* and *Scientific American Mind* since 1999 in this collection.

30 illustrations.

Cloth • 300 pp • ISBN-13: 978-1-932594-22-5 • $25.00

CEREBRUM 2007: Emerging Ideas in Brain Science

Cynthia A. Read, Editor • Foreword by Bruce McEwen, Ph.D.

Prominent scientists and other thinkers explain, applaud, and protest new ideas arising from discoveries about the brain in this first yearly anthology from *Cerebrum*'s Web journal for inquisitive general readers.

Visit *Cerebrum* online at www.dana.org/news/cerebrum.

10 illustrations.

Paper • 243 pp • ISBN-13: 978-1-932594-24-9 • $14.95

MIND WARS: Brain Research and National Defense

Jonathan D. Moreno, Ph.D.

A leading ethicist examines national security agencies' work on defense applications of brain science, and the ethical issues to consider.

Cloth • 210 pp • ISBN-10: 1-932594-16-7 • $23.95

THE DANA GUIDE TO BRAIN HEALTH:
A Practical Family Reference from Medical Experts (with CD-ROM)

Floyd E. Bloom, M.D., M. Flint Beal, M.D., and David J. Kupfer, M.D., Editors

Foreword by William Safire

A complete, authoritative, family-friendly guide to the brain's development, health, and disorders. 16 full-color pages and more than 200 black-and-white drawings.

Paper (with CD-ROM) • 733 pp • ISBN-10: 1-932594-10-8 • $25.00

THE CREATING BRAIN: The Neuroscience of Genius

Nancy C. Andreasen, M.D., Ph.D.

A noted psychiatrist and best-selling author explores how the brain achieves creative breakthroughs, including questions such as how creative people are different and the difference between genius and intelligence.

33 illustrations/photos.

Cloth • 197 pp • ISBN-10: 1-932594-07-8 • $23.95

THE ETIIICAL BRAIN

Michael S. Gazzaniga, Ph.D.

Explores how the lessons of neuroscience help resolve today's ethical dilemmas, ranging from when life begins to free will and criminal responsibility.

Cloth • 201 pp • ISBN-10: 1-932594-01-9 • $25.00

A GOOD START IN LIFE:
Understanding Your Child's Brain and Behavior from Birth to Age 6

Norbert Herschkowitz, M.D., and Elinore Chapman Herschkowitz

The authors show how brain development shapes a child's personality and behavior, discussing appropriate rule-setting, the child's moral sense, temperament, language, playing, aggression, impulse control, and empathy.

13 illustrations.

Cloth • 283 pp • ISBN-10: 0-309-07639-0 • $22.95
Paper (Updated with new material) • 312 pp • ISBN-10: 0-9723830-5-0 • $13.95

BACK FROM THE BRINK:
How Crises Spur Doctors to New Discoveries about the Brain

Edward J. Sylvester

In two academic medical centers, Columbia's New York Presbyterian and Johns Hopkins Medical Institutions, a new breed of doctor, the neurointensivist, saves patients with life-threatening brain injuries.

16 illustrations/photos.

Cloth • 296 pp • ISBN-10: 0-9723830-4-2 • $25.00

THE BARD ON THE BRAIN: Understanding the Mind
Through the Art of Shakespeare and the Science of Brain Imaging

Paul M. Matthews, M.D., and Jeffrey McQuain, Ph D. • Foreword by Diane Ackerman

Explores the beauty and mystery of the human mind and the workings of the brain, following the path the Bard pointed out in 35 of the most famous speeches from his plays.

100 illustrations.

Cloth • 248 pp • ISBN-10: 0-9723830-2-6 • $35.00

STRIKING BACK AT STROKE: A Doctor-Patient Journal

Cleo Hutton and Louis R. Caplan, M.D.

A personal account, with medical guidance from a leading neurologist, for anyone enduring the changes that a stroke can bring to a life, a family, and a sense of self.

15 illustrations.

Cloth • 240 pp • ISBN-10: 0-9723830-1-8 • $27.00

UNDERSTANDING DEPRESSION:
What We Know and What You Can Do About It

J. Raymond DePaulo, Jr., M.D., and Leslie Alan Horvitz

Foreword by Kay Redfield Jamison, Ph.D.

What depression is, who gets it and why, what happens in the brain, troubles that come with the illness, and the treatments that work.

Cloth • 304 pp • ISBN-10: 0-471-39552-8 • $24.95
Paper • 296 pp • ISBN-10: 0-471-43030-7 • $14.95

KEEP YOUR BRAIN YOUNG:
The Complete Guide to Physical and Emotional Health and Longevity

Guy M. McKhann, M.D., and Marilyn Albert, Ph.D.

Every aspect of aging and the brain: changes in memory, nutrition, mood, sleep, and sex, as well as the later problems in alcohol use, vision, hearing, movement, and balance.

Cloth • 304 pp • ISBN-10: 0-471-40792-5 • $24.95

Paper • 304 pp • ISBN-10: 0-471-43028-5 • $15.95

THE END OF STRESS AS WE KNOW IT

Bruce S. McEwen, Ph.D., with Elizabeth Norton Lasley • Foreword by Robert Sapolsky

How brain and body work under stress and how it is possible to avoid its debilitating effects.

Cloth • 239 pp • ISBN-10: 0-309-07640-4 • $27.95

Paper • 262 pp • ISBN-10: 0-309-09121-7 • $19.95

IN SEARCH OF THE LOST CORD:
Solving the Mystery of Spinal Cord Regeneration

Luba Vikhanski

The story of the scientists and science involved in the international scientific race to find ways to repair the damaged spinal cord and restore movement.

21 photos; 12 illustrations.

Cloth • 269 pp • ISBN-10: 0-309-07437-1 • $27.95

THE SECRET LIFE OF THE BRAIN

Richard Restak, M.D. • Foreword by David Grubin

Companion book to the PBS series of the same name, exploring recent discoveries about the brain from infancy through old age.

Cloth • 201 pp • ISBN-10: 0-309-07435-5 • $35.00

THE LONGEVITY STRATEGY:
How to Live to 100 Using the Brain-Body Connection

David Mahoney and Richard Restak, M.D. • Foreword by William Safire

Advice on the brain and aging well.

Cloth • 250 pp • ISBN-10: 0-471-24867-3 • $22.95

Paper • 272 pp • ISBN-10: 0-471-32794-8 • $14.95

STATES OF MIND:
New Discoveries About How Our Brains Make Us Who We Are

Roberta Conlan, Editor

Adapted from the Dana/Smithsonian Associates lecture series by eight of the country's top brain scientists, including the 2000 Nobel laureate in medicine, Eric Kandel.

Cloth • 214 pp • ISBN-10: 0-471-29963-4 • $24.95

Paper • 224 pp • ISBN-10: 0-471-39973-6 • $18.95

The Dana Foundation Series on Neuroethics

DEFINING RIGHT AND WRONG IN BRAIN SCIENCE:
Essential Readings in Neuroethics

Walter Glannon, Ph.D., Editor

The fifth volume in The Dana Foundation Series on Neuroethics, this collection marks the five-year anniversary of the first meeting in the field of neuroethics, providing readers with the seminal writings on the past, present, and future ethical issues facing neuroscience and society.

Cloth • 350 pp • ISBN-10: 978-1-932594-25-6 • $15.95

HARD SCIENCE, HARD CHOICES:
Facts, Ethics, and Policies Guiding Brain Science Today

Sandra J. Ackerman, Editor

Top scholars and scientists discuss new and complex medical and social ethics brought about by advances in neuroscience. Based on an invitational meeting co-sponsored by the Library of Congress, the National Institutes of Health, the Columbia University Center for Bioethics, and the Dana Foundation.

Paper • 152 pp • ISBN-10: 1-932594-02-7 • $12.95

NEUROSCIENCE AND THE LAW: Brain, Mind, and the Scales of Justice

Brent Garland, Editor. With commissioned papers by Michael S. Gazzaniga, Ph.D., and Megan S. Steven; Laurence R Tancredi, M.D., J.D.; Henry T. Greely, J.D.; and Stephen J. Morse, J.D., Ph.D.

How discoveries in neuroscience influence criminal and civil justice, based on an invitational meeting of 26 top neuroscientists, legal scholars, attorneys, and state and federal judges convened by the Dana Foundation and the American Association for the Advancement of Science.

Paper • 226 pp • ISBN-10: 1-032594-04-3 • $8.95

BEYOND THERAPY: Biotechnology and the Pursuit of Happiness

A Report of the President's Council on Bioethics

Special Foreword by Leon R. Kass, M.D., Chairman

Introduction by William Safire

Can biotechnology satisfy human desires for better children, superior performance, ageless bodies, and happy souls? This report says these possibilities present us with profound ethical challenges and choices. Includes dissenting commentary by scientist members of the Council.

Paper • 376 pp • ISBN-10: 1-932594-05-1 • $10.95

NEUROETHICS: Mapping the Field. Conference Proceedings

Steven J. Marcus, Editor

Proceedings of the landmark 2002 conference organized by Stanford University and the University of California, San Francisco, and sponsored by the Dana Foundation, at which more than 150 neuroscientists, bioethicists, psychiatrists and psychologists, philosophers, and professors of law and public policy debated the ethical implications of neuroscience research findings.

50 illustrations.

Paper • 367 pp • ISBN-10: 0-9723830-0-X • $10.95

Immunology

RESISTANCE: The Human Struggle Against Infection

Norbert Gualde, M.D., translated by Steven Rendall

Traces the histories of epidemics and the emergence or re-emergence of diseases, illustrating how new global strategies and research of the body's own weapons of immunity can work together to fight tomorrow's inevitable infectious outbreaks.

Cloth • 219 pp • ISBN-10: 1-932594-00-0 • $25.00

FATAL SEQUENCE: The Killer Within

Kevin J. Tracey, M.D.

An easily understood account of the spiral of sepsis, a sometimes fatal crisis that most often affects patients fighting off nonfatal illnesses or injury. Tracey puts the scientific and medical story of sepsis in the context of his battle to save a burned baby, a sensitive telling of cutting-edge science.

Cloth • 231 pp • ISBN-10: 1-932594-06-X • $23.95
Paper • 231 pp • ISBN-10: 1-932594-09-4 • $12.95

Arts Education

A WELL-TEMPERED MIND: Using Music to Help Children Listen and Learn

Peter Perret and Janet Fox • Foreword by Maya Angelou

Five musicians enter elementary school classrooms, helping children learn about music and contributing to both higher enthusiasm and improved academic performance. This charming story gives us a taste of things to come in one of the newest areas of brain research: the effect of music on the brain.

12 illustrations.

Cloth • 225 pp • ISBN-10: 1-932594-03-5 • $22.95
Paper • 225 pp • ISBN-10: 1-932594-08-6 • $12.00

Dana Press also offers several free periodicals dealing with arts education, immunology, and brain science. For more information, please visit www.dana.org/books/press